"And there will be peace on Earth whence all mankind had perished" The Red Monk.

Isam Taher Salih.

Frequmechanic Three (Mechanics of frequencies). ©

"Anyone telling us they understand Quantum Mechanics proves they understood nothing about Q.M."

Neil Bohr (Danish Father Of Q.M)

"It's the way nature locks its final secrets imposing its own <u>LIMITS</u> on our knowledge! Did Einstein knew the deeper meaning of the phase angle(Φ)? before he knew the <u>LIMITS</u> of speed of light (C) ??"

"The fact that nature reveal its secrets only by Quantas is ominous that we not allowed complete knowledge "

The Red Monk.

"While Quantum Mechanic requires the Intellectual Leap of not insisting on the details of classical mechanics! Frequmechanic require the Intellectual Leap to further simplify by taking analysis and measurements relative to real or imaginary Cosmic Flux where physical objects identified Digitally by (Φ) Conceptually difficult although similar but not the same attempts established in the science of Hydraulics. Also it maybe helpful revising Geometrical Units briefly outlined in my book Frequmechanic Five,"

"Freumechanics is NEW branch of science founded entirely by myself Isam.T.Salih. based on the following:"

The Philosophy:

"While Quantum Mechanics picture of the world that of FIZZY balls. Frequmechanic see it as frequencies."

Pick up the dictionary of any language and examine <u>Any</u> word written by man and what do you find ? That each word actually represent physical object (Noun,Adjecive etc)Or physical action (Verb)EVEN the words (GHOST Or U.F.O.) Can have physical interpretations in the sense it describe something either been actually seen or <u>Imagined?</u> its existence influenced by cocktail of psychological and physical factors (Coming From within one self) i.e SIGNALS and signals mean frequencies a physical signal from (Man as part of the universe not the universe part of man.)Then look at the word universe there is no plural for it because the physical universe we know is singular therefore it follows that there must be physical network or connective tissue UNITING it ?

Leaving no (VOIDS)?Gravitation and radiations simply not continuous CONNECTIVE MEDIUMS ..Frequmechanivs will clarify the following point :The Standard Model Circumvent this deficiency (Universal Medium)by the innovative term (DARK MATTER) Although most scientist admit it's just a (FUDGE)!When the great Greek Philosopher Pythagoras established his Theory on Triangle $a^2 + b^2 = C^2$ It made no difference if he had drawn triangle in his own mind ?In the Sand ?Or on paper. Similarly Frequmechanic assumptions remains valid applicable And useful. Whether the Cosmic Flux Exist Physically or not the results established in this book are valid ,useful and proven..

Natural Philosophy (Physics).

The Starting point is Heisenberg uncertainty relations $(\Delta X . \Delta P) \sim h$ (Planck Constant) Where p=Momentum While $h \sim X\ 10^{-27}$ erg. Sec

Frequmechanincs attempt to find Universal Constant analogous to (h) Applicable to both classical MACROSCOPIC world and to others as along the following lines:

Acceding to Q.M. Its not possible to make simultaneous measurements of the position X and the momentum P. This IMPLY that there must be infinitesimal interval of TIME (Δt) between the two measurements:

Now let us consider an observer at distant of light Years/Days/Seconds OR even fractions od a second :From MACROSCOPIC objects attempting to measure their position x and momentum P.

Again we can say that due to the limitations imposed by speed of light SIMILTANEOUS measurements are not possible and therefore

$\Delta x . \Delta P \sim \Delta t \approx$ Nuclear Time $X\ 10^{-24}$ Sec. i.e. Function of time.

$\Delta x . \Delta P \sim F(t) = \Delta t / n.C$ Where C=Speed of light.

.n =Number of observations (Eigen Values)

This is best illustrated by the concept of commutation between two operators X&Y again according to Q.M. due to None-Simultaneous observations:

$Xy - YX \neq 0$ ➔ h(Planck's) In the MICROSCOPIC WORLD

<u>While For Classical Observer Simultaneous Observations Seems To Be Possible Only Because The Following COMMUTATION Appear To Equal Zero When In Facts Its</u>

$$XY-YX = \Delta t / n.C \rightarrow 0$$

Also you need (Search---Land Of Plenty) ?

The Science:

"Again In the MACROSCOPIC world Energy means Frequency"

Frequmechanic (F.M)Is Scientific attempt to crush or condense information given by Vectors ? Scalars? Tensors e.g. Length Velocity energy etc in to the Scalars of Frequencies'without losing their track. Proceeding on the arguments that if frequency can be signature for Energy as with Energy =h.ω?then why it cannot be signature for other physical dimensions ?

Its an Ambitious Theory since the route to simplify proved to be bumpy road with many lengthy or cumbersome calculations.

Riddled with errors most of these errors are deliberate mistakes designed to explore the methods and corners peculiar to Q.M. these are delegated 5to Frequmechanic Two Four And Five.

Frequmechanic Three is divided to three PARTS:

Part One :

deals with Measurements.

Part Two :

with the search for universal constant and:

part three :

is selection of relevant equations from my other FOUR books on Frequmechanic An Intellectual Leap required to accept physical objects represented DIGITLY by absolute number without physical units given by φ(Phase Angle).Relative to some REAL or Imaginary Cosmic Flux.!

INTRODUCTION: _____ The Cosmic Flux.

"Whether the Cosmic Flux Physically exist or not ? Frequmechanic remains Valid and useful Science ."

"Whether the Great Greek Philosopher Drew his triangle in his own mindor in the sand or on paper his Pythagoras theorem still hold true !Similarly until the physical existence of the Cosmic Flus Proven experimentally all the arguments Equations and Results in my five bookson Frequmechanics are valid and useful i.e Frequmechanic remains applicable .theoretical device regardless of the Cosmic Flux!

There is nothing new about my idea of the Cosmic Flux! For centuries scientists toyed with the idea of **AETHER** (Greek Word)Filling our universe!Then came Dirac **Sea Of Electrons**! Followed by the Cosmic Fluid ! etc?Except that my idea argues that if there is any kind of universal physicalmedium then it will encounter every single object in the Universe macroscopic or microscopic with certain frequency ($\omega_{cutting}$)Then in order for this medium to remain **Continuous** It must Rejoin itself with another Frequency ($J\omega_{Gluing}$)Otherwise it will become localized losing itsuniversity. Getting ready to repeat the same with the next object. Such that we have :

Notations And Definitions :

ω : For all Frequencies

$\tilde{\omega}$: All Omega Tilda)For Angular-velocities reason for choosing this unusual symbol to imply that it contain the frequency as well as BASIS VECTORS e

e_c and e_g such that in later texts of these five books on Frequmechanics asit can be treated as frequency or angular velocity depending on the context . where its convenient use made of (Vc&Vg) Instead to avoid confusions .

Ω :For The Eigen Values Of My Master Equations.

C :For Cutting the Cosmic Flux .
.g : For gluing or Re-joining it).

rc ,rg= Are Hypothetical Unit radius Vectors ad - infinitum in the direction by which ωc and ωg are Cutting and Gluing the Cosmic Flux Respectively.

$$R_\omega^2 = r_c^2 + r_g^2$$

As to How to obtain these RADII will be shown later.Thus : $\tilde{\omega}c = \omega.rc$ AND $\tilde{\omega}g = \omega.rg$

$\Phi = \| \tilde{\omega}g / \tilde{\omega}c \|$ Phase Angle.

As to How to obtain these RADII will be shown later.Thus : $\tilde{\omega}c = \omega.rc$ AND $\tilde{\omega}g = \omega.rg$

$= \| \tilde{\omega}g / \tilde{\omega}c \|$ Phase Angle.

A- My Master Equation in Basic Form for frequencies :

$$\omega = \omega_{cutting} \pm J\omega_{Gluin}$$

B- My Master Equation in Basic Form for Angular Velocities :

$$\tilde{\omega}(\text{Omega Tilda}) = (\tilde{\omega}_C) \pm j \cdot (\tilde{\omega}_g)$$

C- My Master Equation in Basic Form for Energy :

$$E_{(photons)} = (E_c) \pm j \cdot (E_g)$$

D- My Master Equation in 3-D Basic Form for Velocity :

$$\Omega^{Tilda} = \tilde{\omega}_c \times \tilde{\omega}_g \sin\theta.$$

$$\Omega \cdot R = \omega_c \cdot r_c \times / \omega_g \cdot r_g \cdot \sin\theta. \quad : \quad R = r_c \times r_g \cdot \sin\theta.$$

E - In Tensorial Form: Vspaace ➜ V* space :

$$\Omega_R{}^{\mu}{}_v = (\tilde{\omega}_C)^{\mu} \cdot j \cdot (\tilde{\omega}_g)_v \cdot \sin\theta.$$

F - My Master Equation For The Universal Constant (H):

$$H\mu \; \acute{U} \; H\nu \to h \quad (\; \acute{U} = \text{for intersection}$$

G- My Master Equation For The Destructibility Constant (D).

$$\int e^{iDt} f(t) dt = e^{iDt} f(D) \, dD$$

$$t_1\int^t F_{(Fascist)} \cdot dt = t_1\int^t F_{(Colonialist)} \cdot dt$$

You Also Need (Search---Comparability Index -Z)?

H-The Isamic* Delta.

$$\Phi = \omega g/\omega c = \dot{\omega}g/\dot{\omega}c = Vg/Vc.$$

At : $\Omega=0$ $\Phi_\omega = -j.$ $j=\sqrt{-1}.$

$(J)^n = 1$ At $n=4$ $\Phi_{(V)}= 1$

Thus $(\Phi)^n$ Is normalized at n= 4,16....etc ?

(See ----Master Graph)?

In the three languages ARAIMIC?ARABIC?and HEBREW? The NOUN Isam (My Name)Means SELF-MADE or The Protector also Thee defender.

The ADJECTIV (ISAMIC) prompted by paranoia over copy rights than any EGO-istic/ism.

Frequency (ω) and Angular Velocity ($\dot{\omega}$).

To convert (ω) → ($\dot{\omega}$) We need Unit Radii Vectors :

(r_c) and (r_g) Analogous to the basis vectors (e_1 and e_2).

These are DETERMINED by the DIFFERENTIAL GEOMETRY

Of the INTERACTION between the Cosmic Flux and any object in the universe. Not necessarily the geometry of the object itself but its INTERACTION with the Cosmic Flux.

More illustrations on this METHOD can be found in Frequmichanic -Five and the very idea how(ω)Made to imply the existence of unit vector simplify many calculations ?

Frequmechanics In Matrix Form:

$$\tilde{\varphi} = \begin{Vmatrix} \omega c & \omega g \\ J.r & r \end{Vmatrix} \quad \begin{Vmatrix} J.r. \\ \\ \\ \end{Vmatrix} \text{------- r.}$$

For complete Theory In Matrix Form See Frequmechanic-5)? Let λ =Eigen Value.

$$\lambda . I = \begin{bmatrix} \omega c & \omega g \\ J.rg & rc \end{bmatrix}$$

$$\begin{bmatrix} \lambda - \omega c & \omega g \\ J.rg & \lambda - rc \end{bmatrix} = [0]$$

After expanding this matrix we get the following quadratic equation Let :

$\lambda - \omega c$ = m And $\lambda - rc$ = n Giving:

. mn − J.Vg = 0

But mn = λ^2 + Vc− −$\lambda(\omega c + rc)$

$J.Vg = \lambda^2 + \tilde{\omega}c - \lambda(\omega c + rc)$

$\lambda^2 - \lambda(\omega c + rc) + Vc - Vg = 0$

$\lambda = +(\omega c + rc)/2 \pm \sqrt{[(\omega c + rc)^2 - 4(Vc - Vg)]}/2$

At this juncture we need to recall by definition we said the two operations of cutting and gluing the cosmic flux must follow each other`s very closely otherwise its meaningless to talk about this process therefore both frequencies can have any value each as long as the difference remains inoffensively small by comparison . since the energy required by the Cosmic Flux to cut any object and to glue itself is function of the difference in Frequencies?

Also Depending on the length (L)of the process.

$E = h.\omega = F.L$ (Discrete)   Or

Impulse = $\int$ Force .dt.   (Continuous)

While the angular velocities are Function of the force (F)Only :

The nearest and best illustration is the analogy with the ATTACK ANGLE in AERODYNAMICS. Where the FORCE on the wing is FUNCTION OF the DIFFERENCE between the two pressures otherwise the airplane suffers STALL If this difference →0

Therefore the two velocities $\tilde{\omega}c$ & $\tilde{\omega}g$ Must stay close or the process becomes DEGENERATIVE.

This trailing behind each other is maintained by the fact that the force or the *Pressure of the medium be it AIR or cosmic flux is FUNCTION of FORCE such that the length of the two (rc &rg) Varies according to the size and Geometrical shape of the object

(Time taken if we consider the Impulse route)?

Rc &rg Adjusted to maintain the two ENERGIES (Ec & Eg) close Regardless of the size of the two frequencies. At equilibrium $\Omega r = 0$ Therefore :

$$\lambda = +(\omega c + rc)/2 \pm \sqrt{[(\omega c + rc)^2 - 4(Vc - Vg)]}/2$$

$(\tilde{\omega}c - \tilde{\omega}g) = 0$ Keeping in mind $\tilde{\omega}g = Vg = rg \cdot \omega g$.

$$\lambda = +(\omega c + rc)/2 \pm \sqrt{[(Vc + rc)^2]}/2 \rightarrow$$
$$\lambda = +(\omega c + rc)$$

OR:

$\lambda = 0$.

Now we need to pause ai the following SUBTLE but VERY IMPORTANT POINTS because they constitute part of the proof to my Theory on Frequmechanics.

1-The fact that ($\Omega.r = 0$)when the Eigen value :

($\lambda = 0$).

The significance of this point illustrated by asking which of the following two demands deep delicate reflections?

(a) $\lambda = (\omega c + rc)$ (b) $\lambda = 0$.

The answer is (b)Because :

(λ) = Scalar(ωc) + Vector (rc)

While Ωr Is Vector (Angular Velocity) i.e.

Scalar +Vector = Scalar X Unit Vector

If the Left side =0 then only the Frequency can vanish in the right side not the unit vector since its DIRECTIONAL. So what is so special about:(b) =0 become clear from :

(a) where $\lambda = rc + \omega c$ is meaningless for How can frequency =rc at $\lambda=0$? Implying that at (b) λ must have other value:

.**2**-Throughout the five books I written on Frequmechanics Equilibrium was defined (Not when the two Frequencies are EQUAL?)
But only at

$$J = \omega c/\omega g = 1/\Phi$$

3-This can be seen by SQUARING or raising both sides of the above equality to DISCRETE powers of (n) (There will be more on this (n) in Frequmechanic Five) then multiplying each side of the above equality with mass (m) and by equal but opposite in direction BASIS VECTOR: We obtain equal ACCELERATIONS thus EQUILIBRUIM of the acting FORCES.

4-By definition the Cosmic Flux encounter INFINTE numbers of objects Macroscopic or Microscopic therefore its reasonable to assume that it operate at INFINTE levels (NOT VALUES) of energy such that:

$$\text{LEVEL} = S \times \text{Values}.$$

5- Now back to the matrices when :

$$\begin{bmatrix} m & \omega g \\ J.rg & n \end{bmatrix} \begin{bmatrix} X \\ Y \end{bmatrix} = \begin{bmatrix} 0 \\ 0 \end{bmatrix}$$

Applying Crammer`s rule: Row2 ➔ 4R2 -R1:

$$\begin{bmatrix} m & \omega g \\ 4jrg-m & 4n-\omega g \end{bmatrix} \begin{bmatrix} X \\ Y \end{bmatrix} = \begin{bmatrix} 0 \\ 0 \end{bmatrix}$$

[m.x +ωg.y] =0 ➔   x/y =-ωg/m

[4xJ rg -mx + 4yn-yωg] =0➔

x/y = (- 4n +ωg)/ (4J rg -m) =-ωg/ m

x/y = (4n/ωg +1) /(4rg/ωg -1)

Substituting back for n= λ -rg

x/y = (4λ/ωg − 4rg/ωg+1) /(4rg/ωg -1)

➔   X =(4λ/ωg − 4rg/ωg+1)

Y= 4rg/ωg -1)

Let : 4rg/ωg+1) =K And (4rg/ωg -1)=L

At λ =0

X/y = K/L

The significance of this result that the vector is now expressed only with one freanduency and its unit basis vecyor State S(g) Gluing. $S_{(g)} \cdot [K]$

[L]

Therefore, the importance of this result is that in PRACTICE although we can detect TWO FREQUENCIES but ONLY ONE can be measured.

AGAIN AT: $\lambda = \omega c + rc$

$\lambda = +(\omega c + rc)/2 \pm \sqrt{[(\omega c + rc)^2 - 4(V_c - V_g)]}/2 \rightarrow$

$(\omega c + rc) = \pm \sqrt{[(\omega + rc)^2 - 4(V_c - V_g)^2}$

Let $u = (\omega c + rc)$ AND $(V_c - V_g) = P$ Then at $\lambda = 0$

$U^2 = \pm [U^2 - 4(P)^2] \rightarrow \sqrt{4(P)^2} = 0$

OR $2 \cdot U^2 = 4(P)^2$ (Clearly by expanding brackets only these contain terms for velocities and accelerations) →

$U = \pm \sqrt{2} P$ Giving the following FOUR vectors:

UP = [U]   [1]   [1]   [0]   [0]
     [P]   [+√2]   [−√2]   [+1]   [−1]

More dimensions can be considered it becomes cumbersome: The question now when this process by the Cosmic Flux becomes DEGENERATIVE?

Consider The Following arrangements Of Matrices :

$$A = \begin{bmatrix} +1 & -1 \\ 1+\sqrt{2} & 1-\sqrt{2} \end{bmatrix}$$

$$B = \begin{bmatrix} -1 & +1 \\ 1+\sqrt{2} & 1-\sqrt{2} \end{bmatrix}$$

$$C = \begin{bmatrix} 1 & -\sqrt{2} \\ -1 & 1+\sqrt{2} \end{bmatrix}$$

$$D = \begin{bmatrix} 0 & 0 \\ 1 & -1 \end{bmatrix} \quad E = \begin{bmatrix} +1 & 0 \\ 0 & -1 \end{bmatrix}$$

$$F = \begin{bmatrix} 0 & -1 \\ +1 & 0 \end{bmatrix} \quad G = \begin{bmatrix} 0 & 1 \\ 1 & 0 \end{bmatrix} \quad H = \begin{bmatrix} 1 & 1 \\ 1 & -1 \end{bmatrix}$$

The determinants of :

.a= 2 . b= -2 c= +2√2 d= 0. e=-1 f= +1.

Therefore only(D) degenerative: i.e. Pr=1/6.
(Search---<u>Spherical Ensembles And Equipotetials.And Unit Sphere V)?</u>

Quantum Logic Of Frequmechanics:

We can construct Quantum Logic from the above matrices obtained by Frequmechanic proceeding as in the following EXAMPES:
Translated by the Clifford Notations: Identity and global phase:

$e^{i\Phi} = 1$ For any Φ And $e^{i\sigma} I\Psi> \otimes I\Phi> = e^{i\Phi} I\Psi> \otimes I\Phi>$
Matrix D= The Null (Φ)
Matrix F= the identity (I).
Examples:
1- X not σx → ⊗ OR ⊕ See matrix G above.
2- Y, σy → [Y] See matrix F above.
3- Z, σz → [Z] See matrix E above.

Then proceeding as follows:

Hermitian [H]→ $1/\sqrt{2}$ [1   1]
              [1 -1]
Can be derived from above matrices as follows:

$1/\sqrt{2}$ [$\sqrt{2}$ $\sqrt{2}$]
        [$\sqrt{2}$ -$\sqrt{2}$]

Which is equivalent to in terms of the above established matrices And determinants :
$$a/c [H].$$
AND SO FORTH AND SO ON ; Thus a system and basic framework been established for Frequmechanic.

Another Way For Understanding Frequencies.

"Readers seeking concrete confirmation and verification of all my equations of Frequmechanics are referred to brilliant book :

(Physical principles of quantum theory) By Werner Heisenberg b Readers will find The following <u>THREE</u> fundamental agreements :

1-<u>The Rydberg-Ritz. Combination Theory.</u> $\omega_{n.m} = \omega_{n,k} + \omega_{k.m}$

The main point in this theory which are two terms : T1 and T2 are very close:

$$\omega_{n.m} = \omega_n \pm \omega_m.$$

Its of the same idea as my postulate that ωcutting and ωGluing

Are trailing behind each other closely! Where the two terms T1 and T2 Arranged in one dimensional matrice ARRAY(n,m) ?T The ATOMIC FREQUENCIES form Two dimensional array(k,n,m)!Also we need to keep in mind when discussing the Universal Constant as multiples of (h)In my Frequmechanic -Five (under The Title In Search Of My Universal Constant) We Need to remember The following from the book by Heisenberg:

$J_k = n_k.h$ Where (J_k) Are known as ACTION VARIABLES !And the (n_k)Are integers for
2- <u>Bohr Frequency Condition</u> .(k=(1.2.3......f)

$\omega_{n.m} = 1/h$ (En - Em) Where E and $\omega_{n.m}$ Are the characteristic frequencies and energies The analogue to the classical: $\omega_k = \partial H/\partial J_k$

The canonical Conjugate known as Action-Angle variables (J_k and w_k) Defined as :

H the Hamiltonian depend only on (J_k)While ω_k related to the Frequency by   :

$W_k = \omega_k + \beta_k.$ Where β_k=Constants(Plr). $J_k = \partial H/\partial w_k$ $W_k = \partial H/\partial J_k = \omega_k$.

3-The Action-Angle Variable :

Also the reader is referred to how the same book explore the possibilities of expressing frequencies by alternative co-ordinates and how the Amplitude can be explicitly TIME independent with Fourier series and its CONJUGATE (J)of harmonics. For any multiply periodic ?Also HOW quantum frequencies obey the Rydberg - Ritz principle This is important point to keep in mind when my book Frequmechanic-Five will establish the verifying result by obtaining quantum (Discreet) Numbers from my equations listed above at ($\tilde{\omega}C$) =($\tilde{\omega}g$)?And this is the reason for quoting this book by Heisenberg for the reader to refer to as it shows very clearly that in the stationary states (i.e. When ($\tilde{\omega}C$ =($\tilde{\omega}g$ according to Frewumechanics) It was proven Empirically that the INTENSITIES calculated CLASSICALY by Fourier amplitudes is in FULL <u>AGREEMENT</u> with the intensities of the stationary state when the quantum integers are very large.

Justifications:

"One of the many justifications for Frequmechanics quoted in my Five Volumes on Frequmechanics Is the following: To talk about fraction of cycle is meaningless in our present knowledge and human perceptions thus frequencies can only be wholele integers. Whole integers can be cyclic philosophically interpreted as our perception is intrinsically discrete.

Just ask : Why the METRIC that stem defeated the fractional so called Imperial system??"

The book by Heisenberg brilliantly draw s our attention to concrete analogues between classical and quantum theories. Such

1- By taking the POISSON BRACKETS over to Quantum Theory introducing the ACTION-ANGLE Variables :

2- By splitting the definition of continuous classical functions by **PARTIAL DIFFERENTIATION** to the quantum case ! Where :

$\partial Jk \rightarrow \propto \kappa \cdot h$ & $H \rightarrow H(n1,n2,n3.....nk$ - k From : $H(n1,n2,n3....nf.)$.

All of these are necessary to understand the bases of my Frequmechanics since Here we need to quote again two of the bases of Frequmechanics:

1- The two frequencies ωcutting and ωGluingAre trailing behind each other closely!

Thus the Rydberg-Rits combination Rule. Was Satisfied.

2- Although Frequmechanics dealing with two frequencies of cutting and gluing its bridged by CONTINOUS Physical feedback.(We called Cosmic Flux)

However, my Frequmechanic -Five book offers simpler but less convincing route by the following equation:

$$\Omega_R{}^{\mu \cdot \nu} = (\tilde{\omega}_c)^\mu \cdot J(\tilde{\omega}_g)^\nu \; \text{Sin} \; \Phi \qquad \text{-EQN-1}$$

<u>EXERCISE:</u>

The above Equation - One is Hybrid equation contain Quantum and Classical

Split and Specify which Terms are Discrete and which is Continuous??

The most relevant part of the book is how Heisenberg remind us of the origin of Planck's Constant $(h/2\pi)$ From the non-commutative part of Pioson bracket relationship for MICROSCOPIC objects This is in line with what Frequmechanic trying to establish under the title (In search for my universal constant) (Search---Land Of The Plenty) ?

$$N/Z_{(A)} = K = 1/500 \times 10^{+27}$$

Basica stating that if we can (In theory at least)Put astronomical numbers (N) of MACROSCOPIC particles e.g exact copies of sand grains in COMPACT SPACE and observe them from astronomical distance (The inverse of Atomic distances)e.g (one centimeter / 0.2×10^{-13} ≈of the order ($2 \times 10^{+14}$ CM.) Then we should be able to observe Quantum Fields(e.g Waves) Such (View)Can offer explanations to the many unexplained phenomena in the universe pointed out in the Five books of Frequmechanics! The section titled (LAND OF THE PLENTY) Establishes concrete equations between the minimum(N)and Distances wwhen Quantum Effects can be detected ?Now back to the Poison Brackets):

Let C & g As defined in Frequmechanics.

Let n,m, K, α_k, β_k. $W_k = C\omega_k + \beta_k$.

And(n-m = $\alpha - \beta$.) .

As defined in Heisenberg book . Now :

Let : R= n- α , S= n- β, T= n- α - β., C=n , g=m.

At $\alpha, \beta \ll n$.

$i.h[x(C,R)-x(S,T)].Y(R,T) - [Y(g,S) - Y(R,T).(S,T)$.

$\rightarrow h\beta \; \partial x_\alpha (n)/\partial s$

$Y(R,T) \approx \{1/2\pi . i.\beta \} [Y_\beta (R)] / \partial w \rightarrow 1/2\pi. i.\beta [Y_\beta (n)] / \partial w$

Notice How the C,g,and h Disappeared from the second part of the βexpression? Clearly the **MATHEMATICAL** symmetry is strikingly.

But is it enough ?: Attempts to answer this question will be made in this book further in Frequmechanic -Five.

Classical $|x,y| = xy-yx = 0$ In Both cases.

Quantum $|x,y| = xy-yx = i.h$ At $n=m$.

Frequmechanical $|x,y| = xy-yx = i. H^2 .(\phi)^{C,g.}$ At $C=g$.

Let us express the Angular <u>Velocities</u> $\tilde{\omega}$ in terms of generalized coordinates as follows: $\tilde{\omega}c = q^c{}_\alpha$, $\tilde{\omega}g = q^\beta{}_g$

By definition $dx^\alpha/dx^\beta = H$ (Where H is Time-Space Coordinate):

$\tilde{\omega}c . \tilde{\omega}g = q^c{}_\alpha . q^\beta{}_g = q^c{}_g$

BY Substituting the following in to the above Poison Brackets

And Collecting All (Γ)Terms : $.dq^c = \Gamma^c{}_{g,v} q^v dX^g \rightarrow$

$$(\phi)^{C,g.} = (1/H) . q^c{}_g$$

Parodoxes From Relativity ForThe Michilson-Morley Experiment? The Aeather And The Cosmic Flux.

Consider the following simple mechanical thought experiment :
Two trains moving towards each otherb and each of their own speedometer recording their velocities respe as (V1)And (V2)Respectively. Then imagine there is an observer (C) right up there above them in the sky measuring their COMMON SPEED (V)
He measures the distance (L)between them (At any Certain moment Then the time left before collision = L/V   Where V=V1 +V2.
Again if the two trains happened to be running away from each other the same apply Time =L/V Where L now is ANY constant distant between them (Until they stop).In both cases above the two trains were running in Opposite Directions !However the above argument do not apply if they were running in the Same Direction since there is no way for observer (C) To determine any COMMON SPEED all he has is the constant length (L) maintained by two constant speeds. Its NONESENSICAL for him to talk of any such (V)Since there is no way for him to Make any measurements as he did above when the trains were running in opposite direction . i.e. the reason such calculations are possible only because relativistic calculations cancel each other's BY SYMMETRY when they motions are opposite each other in which case its not possible to measure the DRIFT of the AETHER or the COSMIC FLUX.Erroneously text books on relativity ignore this point ?So we delegate this riddle to Quantum Mechanics and Frequmechanics??

Two derivative operators on Scalar Field (^▼).

Frequency Ω Is essentially complex number   !

but in frequmechanics both (Ω & Φ) Can represent physical objects. Whether the cosmic flux physically exist or not all analysis in FREQUMECHANICS are GROUP (G)THEORIC Valid applicable and useful : G(gluing)/G (Cutting) = I(The Identity). V➔ V* Vector Space ➔Covectors: ῶc ➔ ῶg The assumption of:

$$E = h.\Omega$$

can be valid for matter as well as for radiation by computing the GROUP velocity of the wave packet representing Non-Relativistic particle with : Energy E mass m and momentum p where :

$$p = h/\lambda \, !$$

GROUP velocity is the velocity of the center of the wave packet. <u>EMITTON</u> is The suggested particle of Frequmechanic it will be introduced in Frequmechanic -Five including the full calculations of its specifications e.g wave Length SPIN. Etc. Comparisons are made with photons by approximating the geometrical path to that of photon in vacuum.

Q.M.

Whenever in later discussions the Quantum-Picture to my theory is presented OMEGA-Capital version treated as the EIGEN VALUE of my master Equations:

$$(\Omega) \cdot I = \begin{bmatrix} \dot\omega_c & \dot\omega_g \\ J \cdot R_g & R_c \end{bmatrix}$$

Where (I) is the IDENTITY MATRIX giving the following quadratic equation:

$(\Omega)^2 - \dot\omega_c \cdot R_c + \dot\omega_g \cdot R_g = 0 \rightarrow (\omega \cdot R)^2 = (\dot\omega_c \cdot R_c - \dot\omega_g \cdot R_g)$

$\rightarrow (r_c^2 + r_g^2) \omega^2 = (\dot\omega_c \cdot R_c - \dot\omega_g \cdot R_g)$

$\omega^2 = (\dot\omega_c \cdot R_c - \dot\omega_g \cdot R_g) / R^2$

$\omega = \pm \sqrt{(\dot\omega_c \cdot R_c - \dot\omega_g \cdot R_g)} / R$ Giving:

$\omega \cdot R = \pm \sqrt{(\dot\omega_c \cdot R_c - \dot\omega_g \cdot R_g)}$ OR $\tilde\omega = \pm \sqrt{(\dot\omega_c \cdot R_c - \dot\omega_g \cdot R_g)}$

EXERCISE:

Q- What is the significance of the above result?

A- Theory is valid for normalizes Angular Velocity $(\tilde\omega)^2$

JUSTIFICATIONS.

1-

If the word (ONE) in the phrase (one universe) to be viable there MUST be (ONE) Tissue connecting this universe hereafter we call it the COSMIC FLUX.(C.F)

2- Its reasonable to assume that this C.F Will be encountering each object in the universe macroscopic or microscopic thus cutting this C.F. with frequency ($\omega_{cutting}$);

3-

Therefore it follows that for this C.F. To remain universal it MUST Rjpoint itself at frequency (ω_{gluing}) Otherwise it will be LOCALIZED not universalized.

4- According to Q.M. Its not possible to measure these two frequencies simultaneously but we can measure (Ω) Where this (Ω) Represent the following possibilities.

(Ω)1 = (ω_c) + (ω_g) Classical

(Ω)2 = (ω_c) - (ω_g) Classical

(Ω)3 = (ω_c) + J. (ω_g) Q.M

(Ω)4 = (ω_c) - J. (ω_g) Q.m.

TESTING BY INSPECTIONS:

Our tool for testing which of the above equation is viable by probing the Equilibrium state when $(\omega_c) = (\omega_g) = \omega$ Giving: $(\Omega)1 = 2(\omega)$

$(\Omega)^2 1 = 4.(\omega)$

$(\Omega)2 = (\omega) - (\omega) = 0 \quad (\Omega)^2 2 = 0.$

$(\Omega)3 = (\omega_c) + J.(\omega_g)$ At $\omega_c = \omega_g$

$(\Omega)3 = \omega (1+ J.)$

$(\Omega)^2 3 = (\omega)^2 + (J\omega)^2 + 2.J(\omega)^2 = 2.J(\omega)^2$

$(\Omega)4 = \omega (1 - J.)$ At $\omega_c = \omega_g$

$(\Omega)^2 4 = (\omega)^2 - (J\omega)^2 - 2.J(\omega)^2 = 2(\omega)^2 (1-J)$

$(\Omega)^2 4 + (\Omega)^2 3 = 2(\omega)^2$ i.e NORMALISED) and REAL

Therefore the best candidate is: $(\Omega)4 = (\omega_c) \pm J.(\omega_g)$

Sinc at equilibrium $(\omega_c) = (\omega_g)$. *Meaning by splitting the master frequency* (Ω) *In to: REAL* (ω_c) *And IMAGINARY parts* $(J.\omega_g)$ *Does not effects the result* $2(\omega)^2$ *Which remains REAL.*

Exercise :
Which of the above relations falls in the First second third or fourth quadrant ointhe ARGAND diagram?

ANSWER:

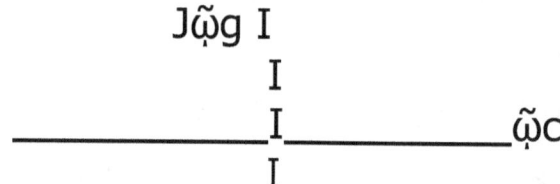

Whether the C.F. Is cutting and Gluing the object by $(\Omega)3$ Or $(\Omega)4$ Above or below the Real axis ?Right or left of the Imaginary Axis the result is REAL and the same $=2(\omega)^2$

THE PROBLEM OF DIRECTIONS:

"it's the problem of finding mechanism where $\tilde{\omega}c$ & $\tilde{\omega}$ are ORTHOGONAL as to each other inside plane normal to the direction of motion."

1- For free falling particle under gravity acceleration (g) The path of such particle will be vertical. Thus The Emitons (The sugested particle of Freumechanic) in straight line approximation can be compared to that of Photons,   h.w = E

2- As it had been stated in all my five frequmechanoc books Of the two frequencies cutting and rejoining the C.F. need to be orthogonal to this path thus (θ)<u>Need to be as near as possible to</u> (п/2). Defining hypersurface ?

3- Again as stated in all my Five Frequmechanic books that the two Frequencies (ωc and ωg) Need to be trailing behind each others very closely i.e. by $\delta\omega$ and the lag or phase angle

$$\Phi = \| \omega g / \omega c \| \approx 1 .$$

These three conditions best illustrated by the following three dimensional diagram:

3D(XYZ, θ)

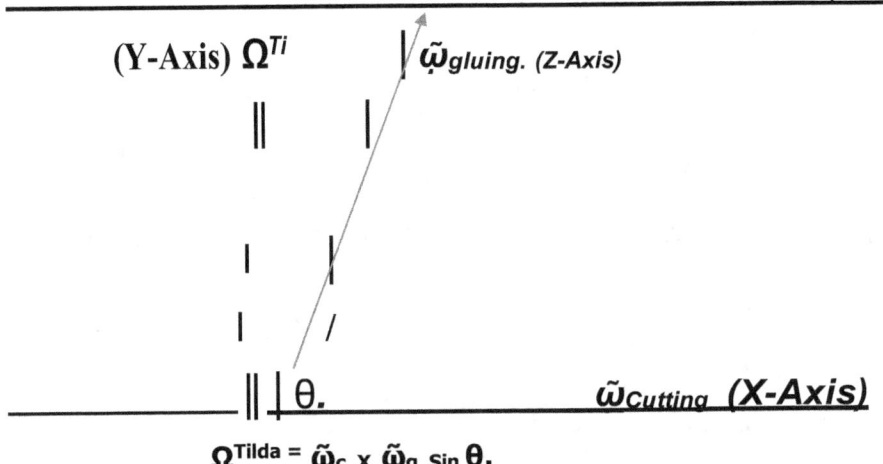

$\Omega^{Tilda} = \tilde{\omega}_c \times \tilde{\omega}_g \sin\theta.$

$\Omega.R = \omega_c . r_c \times \omega_g . r_g . \sin\theta.$ But:

$R = r_c \times r_g . \sin\theta.$

Therefore the unit radius can be removed from both sides of the equation giving And the most important result that hereafter angular velocities by implication can be treated as Frequencies by implications in differentiating with respect to time since the Radii remains constants. $\Omega = \omega_c \times \omega_g$ Differentiating:

$d\Omega/dt = \omega_c \times d(\omega_g)/dt + \omega_g \, d\omega_c/dt$

Nultiplying through by dt and Integrating:

$\Omega = \omega_c . \omega_g + \omega_g , \omega_c + \text{Constant} = 2\omega_c . \omega_g + \text{Constant}$

Applying by inspection my testing method at EQUILIBIRIUM :

$$\omega_c = \omega_g$$

½ $\Omega = \|\omega_c\|^2 +$ Constant OR;   ½ $\Omega = \|\omega_g\|^2 +$ Constant

¼ Ω = ($\|\omega c\|^2$ +Constant)($\|\omega g\|^2$ +Constant)

At equilibrium: $\Omega^2 = 4(\|\omega\|^2 + K)2$

$$\Omega^2 = 2(\|\omega\|^2 + K)$$

Ths is Quadratic equation : Let the constant of integration =g (Gravity acceleration. $\|\omega c\|^2$ = C and multiply through by R

$\Omega^2 \cdot R = R \cdot C^2 + (g) \cdot R$

$R = g \cdot C + g + (g)^2$

Let C.P =Centripetal acceleration.

.g= (Total Observable (C.P)- Invisible C.P)/R ----------------- EQN-A

Alternatively:

$\Omega \cdot R = \omega c \cdot rc \times /\omega g \cdot rg \cdot \underline{\sin\theta} = \omega c \cdot \omega g \, rc \times rg \cdot \underline{\sin\theta}$

Diiferentiating all terms with respect to t:

$R \cdot d\Omega/dt + \Omega \, dR/dt = \omega c \cdot rc \times rg \cdot \sin\theta \cdot \boldsymbol{d\omega g}/dt$

$+ \omega g \, rc \times rg \cdot \sin\theta \, \boldsymbol{d\omega c}/dt + \omega c \cdot \omega g \, rg \cdot \sin\theta \cdot \boldsymbol{drc}/dt$

$+ \omega c \cdot \omega g \, rc \, \sin \, drg/dt + \omega c \cdot \omega g \, rc \times rg \cdot \cos\underline{\theta \, d\theta/dt.}$

Multiplying throuogh by (dt)Then integrating:

$R \cdot \Omega + \Omega \cdot R = \omega c \cdot rc \times rg \cdot \sin\theta \cdot \omega g$

$+ \omega g \, rc \times rg \cdot \sin\theta \, \omega c + \omega c \cdot \omega g \, rg \cdot \sin\theta \cdot \boldsymbol{rc} + \omega c \cdot \omega g \, rc \, \sin$

$\theta \cdot \boldsymbol{rg} + + \omega c \cdot \omega g \, rc \times rg \cdot \cos\underline{\theta}$ +Constant of integration (g)

Applying The equilibrium test at   ($\omega c = \omega g = +\omega$

$2 R. \Omega = 2 \Omega^{Tilda} = 2\omega^2$ rc X rg. $\sin \theta + 2\omega^2$ rc. rg. $\sin \theta$. $+\omega^2$ rc rg. $\cos \underline{\theta} \cdot \underline{\theta} + g$

$\Omega^{Tild}/(\omega^2 \cdot rc \cdot rg) = {}_{2\sin} \theta. + \cos \underline{\theta} \cdot \underline{\theta} +$ Constant

At $\theta = \pi/2$ rc→g

$r\Omega^{Tild}/(\omega.r)^2 = {}_{2\sin} \pi/2 + \underline{.\theta} \cdot \cos \pi/2 +$ Constant

$\Omega^{Tild}/(\tilde{\omega})^2 = {}_{2\sin} \pi/2 + \underline{.\theta} \cdot \cos \pi/2 +$ Constant

$\Omega^{Tild}/(\tilde{\omega})^2 = 2 +$ Constant $=_{Constant} \Phi$

(Lag or Phase angle in Seconds) $= \Omega^{Tilda}/(\tilde{\omega})^2$ ---------------- EQN=B.

RESULTS:

Cleary it can be seen from EQN A and EQN-B that the acceleration due to gravity (g) and the phase angle (Φ) Are related Organically Physically and Mathematically.

The Limits Of Phase Angle:

"Maximum information obtained when (β) At Minimum."

"Just as nature put limit on speed of light (C) The phase angle (Φ) Had been shown above LIMITED For : $(\Phi)^{-\beta}$ $\beta \neq 0$

What is the Significance of the above equations

Clearly this result proves that EVEN when the two frequencies ωc and ωg | are equals and identical in value and directions there always will be infinitesimal phase or lag angles between them : therefore we have so far <u>JUSTIFIED</u> the splitting of the Frequency($\Omega\_$ In to TWO. <u>Moreover Its First Ever Proof Of The E existence Of Two Frequencies Embedded In One.</u>

Mass And The Poloraization Of Photon:

$\omega \cdot h = \omega c \cdot h \pm J\omega g \cdot h$ h-bar. = Plank constant.

$E_{photons} = (Ec) \pm j \cdot (Eg)$

$\Omega \cdot R = \omega c \cdot rc \times \omega g \cdot rg \cdot \sin\theta$: $R = rc \times rg \cdot \sin\theta$.

$\tilde{\omega}$ (Omega Tilda) $= (\tilde{\omega}C) \pm j \cdot (\tilde{\omega}g)$

$\Omega^{Tilda} = \tilde{\omega}_c \times \tilde{\omega}_g \sin\theta$.

At Equilibrium: $\tilde{\omega}_c = \tilde{\omega}_g$ AND by definition $rc = rg \rightarrow$

$\Omega \cdot R = (\omega \cdot r)^2 \rightarrow \Omega = (\omega)^2 \cdot (r)^2 \cdot \underline{\sin\theta}$.

Energy $= \Omega \cdot h = \{ \frac{1}{2} m (\omega)^2 \cdot r^2 / R \} \sin\theta$.

After multiplying R.H.S. by MASS(m):

Mass (m) of Photon $= 2R/r^2 \cdot (\Omega \cdot h) / (\omega)^2 \cdot \sin\theta$.

Since by definition: $\sin\theta = 1/\sqrt{2}$

Also at equilibrium; $R^2 = 2r^2$ $R = \sqrt{2}$

Since $rc = \pm rg$ = one unit vector $(e) = 1$

$m = 2 \cdot \sqrt{2} / \sqrt{2} \cdot h \cdot (\omega^2)/\Omega$.

The Limits are : $\omega c = \Omega$ OR $\omega g = \Omega$ $\rightarrow$

$m = (1/\Omega) \, 2 \cdot h \cdot C^2 \cdot <C | \omega | g>$

C = Speed of light $\approx 10^9$ Cm/Sec. $h \approx 10^{-27}$ erg.se

Let $<C | \omega | g> / \Omega = P(r)$

$m \approx 2 \cdot 10^{-27}/10^9 \, P \cdot 10^{-9} \cdot P \cdot$ (gm). $\Omega >> \omega$

Regional Constants $(H)_{\mu,\nu}$

The General Argument Proceeding on the assumption that Frequency can be signature for energy as evident from planck constant for photons

$$E = h.\omega$$

Then why not for all other objects microscopic or macroscopic if the right constant a regional constant can be predetermined ?

But this raise another question of how to define the (REGION)?

Let both the region and its constant identify themselves by themselves?

Based on the facts that although (Alpha $(\tilde{\alpha})$ = $10^{-(Minus\ Betta)}$

The range of $\beta = 0 - \infty$ Thus the region will be a MAP from $(\alpha : \beta)$

$\alpha = (1 \rightarrow 0)$ To Betta $(0 \rightarrow \infty$ Such that:

$$(\alpha)\mu = (H)\mu, V\ \beta\nu$$

Reminder :C (For Cutting and g(For gluing).As Dummy Indices.

$$d(\alpha)\mu / d\beta\nu = \Gamma\mu\nu, C \qquad\qquad d(\alpha)\mu / d\beta\nu = \Gamma\mu\nu, g$$

$$\Gamma\mu\nu, C + \Gamma\mu\nu, g = ?$$

Applying the identity: $d\tilde{\omega}C\ \Gamma\ \alpha\ C.\nu, \mu. = d\tilde{\omega}g\ \Gamma\ \alpha g, \mu., \nu.$ Always keeping in mind:C (For Cutting and g(For gluing).

Defining My Universal Constant:

Let there be two very close adjacent regions (C) and (g)

Almost touching each other with the intersection kept Infinitesimally Small So that can be easily classified as microscopic such that the constant h(Planck`s)Is Valid while the two classical constants H μ and H ν will be valid in their respective regions(C and g)Such that:

H μ Ú H ν (U for intersections)➔ h(Planck) ,

One Leg In Planck,s !One Leg Out!!

"The photon travels all the way from distant galaxies with frequency ωp_{photon} unhindered by the C.F. reaching the eye OXIDIZING it whenever ωp Resonate with either frequencies of the Cosmic Flux ωc OR ωg depending if Φ Leading or Lagging(It's the same-the first leg)Thus we can replace ωp with one and only one at a time of the the two C.F frequencies as follow:)!"

$$H\mu \; \acute{U} \; H\nu \rightarrow h \quad (\acute{U} = \text{for intersections})$$

Then multiplying by respective frequency: Ec-Eg = Ep (Photon.)

Similarly Ec = H μ . ωc   AND   Eg = Hν . ωg   GIVING:

Ec/Eg = (H μ . ωc)/(H ν . ωg) BUT (ωc/ ωg)= Φ

Ec/Eg = (H μ /H ν) . Φ . NOW Let (H μ /H ν)= H μ,ν   Therefore:

Ec/Eg =   H μ,ν . Φ NOW:

A-Let ωp = ωc (First Leg)→   Ep/Eg = {(h. ωp)/Eg } OR not AND.

B-Let ωp = ωg (First Leg)→   Ec/Ep = {Ec/(h. ωp) }

By Multiplying :   A x B→ Ec=Eg

If Ec/Eg = H μ,ν . Φ It means when (ωc= ωg)At equilibrium:

And $\Phi \sim 1$.   H μ,ν → 1. OR :

H μ, = H ν = H

Thus the Universality of the constant (H) Is valid for all frequencies.
Contrary to appearance this is Important result since :

H μ.ωc = J.Hv .ωg.   OR   μ.ωp = J.Hv .ωg. OR: H μ.ωc = J.h .ωp.

in case the significant of the above equation is lost there is a need to emphasize that the R.H.S. of the above equations is valid and had been experimentally verified for microscopic particle(The Photon. This imply the L.H.S, describing Macroscopic Particles also valid in turn validating the Regional Constants inside this **Hybrid** Zone with the following conditions:

(H)μ + (HV = (H)μ,V   AND:

PROOF:

Let : (H)μ ≠ (H)v   Rc- Eg = H μ. ωc - Hv. ωg=h.ωp.

At equilibrium(Resonance)   ωc= ωg = ωn.

H μ. ωn - Hv. ωn=h.ωp. → H μ. - Hv. =h.ωp/ωn .

Now the argument is that: If H μ.=Hv. then h.ωp/ωn =0

Since h ≠0 Therefore ∏ ωp/ωn∏=0

Vut we know for sure ωp ≠0 therefore ωn→∞ (Infinity)

And this stand to reason since in strong gravitational fields its **MEANIGLESS** at infinity (Extremely Large Values)to distinguish between two frequencies ωc & ωg Only One Natural frequency ωn make sense.

Thus   H μ. = Hv.=H

Proving once more the universality NOT locality of (H).

DEFINITION:

PARTICLE:
Any object with physical dimensions e.g Mass?Length etc.i.e Not the statistical probability of its existence.

WAVE:
Describe how such physical particle behave? Thus, the mystery of Quantum Tunnelling can be easily understood :Since the particle as we shall see later emits two frequencies ($\Omega = \omega$cutting $\pm$ JωGluing-EQN-1

$J=\sqrt{-n}$ Where n=0,1,2,3 tc .Always keeping in mind:C (For Cutting and g(For gluing).Introducingangular velocity Omega Tilda :
= $\tilde{\omega}.Rc$ AND $\tilde{\omega}g = \tilde{\omega}.Rg$

Where R =Hypothetical Unit Vector ad -infinitum.

$Rc = H\mu/Kc$ AND $Rg = Hv/Kg$ $H\mu$ and Hv Are the Two components of the universal constant: $H\mu . Hv = H\mu v$
Kc and Kg Are the propsgation vector defined as follows:

What is Natural Frequency?

"Evert object in the universe has Natural Frequency"

$\qquad f = \omega / 2\pi \qquad\qquad \omega = \sqrt{k/m}$

Adjusts our formula to the following:
$2\pi . f = \{\sqrt{k/m}\}/$ K=   Constant   m=Mass.

Exercise:
What is the relationship between K and our (H)Discussed above?

wave vector (K):

The terms wave vector (K) and angular wave vector
$$\tilde{\nu} = |\vec{\nu}|$$
have distinct meanings.
Here, the wave vector is denoted by (K) and the angular wavenumber by v
These are related by
$\mathbf{k} = 2\pi\tilde{\nu}$
A sinusoidal traveling wave follows the equation.

$$\psi(\mathbf{r}, t) = A\cos(\mathbf{k} \cdot \mathbf{r} - \omega t + \varphi),$$

ere:
$\mathbf{r}$ = is position,
t = Time,
ψ is a function of r and t describing the disturbance describing the wave (for example, for an ocean wave, ψ would be the excess height of the water, or for a sound wave, ψ would be the excess air pressure).

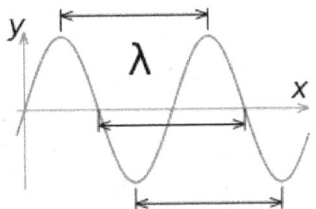

Wavelength of a sine wave, λ, can be measured between any two consecutive points with the same phase, such as between adjacent crests, or troughs, or adjacent zero crossings with the same direction of transit, as shown

A is the amplitude of the wave (the peak magnitude of the oscillation),
- φ is a phase offset,
- ω is the (temporal) angular frequency of the wave, describing how many radians it traverses per unit of time, and related to the period t by the equation $\frac{2\pi}{T}$,
- k is the angular wave vector of the wave, describing how many radians it traverses per unit of distance, and related to the wavelength by the equation $\psi(\mathbf{r}, t)$
= $A\cos(2\pi(\tilde{\nu} \cdot \mathbf{r} - ft) + \varphi)$, where: f is the frquency is the wave vector $(\frac{2\pi}{\lambda})$,

Exercise:
Earlier we said It made no difference which of the two frequencies ωc OR ωg Had its leg first in the <u>Hybrid Zone</u> : How ? and WHY??
Answers :HOW; It if (Φ) Was <u>LEADING or LAGGING.</u>
<u>WHY:</u> It's the familiar fluid flow of no source and no sink probability current density (see Later prob current density)?

Direction of the wave vecto

The direction in which the wave vector points must be distinguished from the "direction of wave propagation". The "direction of wave propagation" is the direction of a wave's energy flow, and the direction that a small wave packet will move, i.e. the direction of the group velocity. For light waves in vacuum, this is also the direction of the Poynting vector. On the other hand, the wave vector points in the direction of phase velocity. In other words, the wave vector points in the normal direction to the surfaces of constant phase, also called wavefronts.

In a lossless isotropic medium such as air, any gas, any liquid, amorphous solids (such as glass), and cubic crystals, the direction of the wavevector is the same as the direction of wave propagation. If the medium is anisotropic, the wave vector in general points in directions other than that of the wave propagation. The wave vector is always perpendicular to surfaces of constant phase.For example, when a wave travels through an anisotropic medium, such as light waves through an asymmetric crystal or sound waves through a sedimentary rock, the wave vector may not point exactly in the direction of wave propagation.

In solid-state physics

In solid-state physics, the "wavevector" (also called k-vector) of an electron or hole in a crystal is the wavevector of its quantum-mechanical wavefunction. These electron waves are not ordinary sinusoidal waves, but they do have a kind of envelope function which is sinusoidal, and the wavevector is defined via that envelope wave, usually using the "physics definition". See Bloch's theorem for further details.

In Special Relativity

A moving wave surface in special relativity may be regarded as a hypersurface (a 3D subspace) in spacetime, formed by all the events passed by the wave surface. A wavetrain (denoted by some variable X) can be regarded as a one-parameter family of such hypersurfaces in spacetime. This variable X is a scalar function of position in spacetime. The derivative of this scalar is a vector that characterizes the wave, the four-wavevector.

$$K^\mu = \left(\frac{\omega}{c}, \vec{k}\right) = \left(\frac{\omega}{c}, \frac{\omega}{v_p}\hat{n}\right) = \left(\frac{2\pi}{cT}, \frac{2\pi\hat{n}}{\lambda}\right)$$

The four-wavevector is a wave four-vector that is defined, in Minkowski coordinates, as:

where the angular frequency $\frac{\omega}{c}$ is the temporal component, and the wavenumber vector $\vec{k}$ is the spatial component.

Alternately, the wavenumber k can be written as the angular frequency ω divided by the phase-velocity vp, or in terms of inverse period T and inverse wavelength λ.

$$K^\mu = \left(\frac{\omega}{c}, k_x, k_y, k_z\right)$$

$$K_\mu = \left(\frac{\omega}{c}, -k_x, -k_y, -k_z\right)$$

When written out explicitly its contravariant and covariant terms are:
In general, the Lorentz scalar magnitude of the wave four-vector is

$m_o = 0$

$$K^\mu K_\mu = \left(\frac{\omega}{c}\right)^2 - k_x^2 - k_y^2 - k_z^2 = \left(\frac{\omega_o}{c}\right)^2 = \left(\frac{m_o c}{\hbar}\right)^2$$

The four-wavevector is null for massless (photonic) particles, where the rest mass mo=0

An example of a null four-wavevector would be a beam of coherent, $v_p \equiv c$ monochromatic light, which has phase-velocity.

$$K^\mu = \left(\frac{\omega}{c}, \vec{k}\right) \quad \left(\frac{\omega}{c}, \frac{\omega}{c}\hat{n}\right)$$

for light-like/null

$$K^\mu = \left(\frac{\omega}{c}, \vec{k}\right) \left(\frac{\omega}{c}, \frac{\omega}{c}\hat{n}\right) \quad \text{\{for light-like/null\}}$$

which would have the following relation between the frequency and the magnitude of the spatial part of the four-wavevector:

$$K^\mu K_\mu \left(\frac{\omega}{c}\right)^2 - k_x^2 - k_y^2 - k_z^2 = 0 \quad \text{\{for light-like/null\}}$$

The four-wavevector is related to the four-momentum as follows:

$$P^\mu = \left(\frac{E}{c}, \vec{p}\right) = \hbar K^\mu = \hbar\left(\frac{\omega}{c}, \vec{k}\right)$$

The four-wavevector is related to the four-frequency as follows:

$$K^\mu = \left(\frac{\omega}{c}, \vec{k}\right) = \left(\frac{2\pi}{c}\right) N^\mu = \left(\frac{2\pi}{c}\right)(\nu, \nu\vec{n})$$

The four-wavevector is related to the four-velocity as follows

$$K^\mu = \left(\frac{\omega}{c}, \vec{k}\right) = \left(\frac{\omega_o}{c^2}\right) U^\mu = \left(\frac{\omega_o}{c^2}\right)\gamma(c, \vec{u})$$

" If we can accept all the arguments that produced :
($E = h.\omega$) For the Microscopic world ? Then it should be equally valid for ($E = H.\omega$) In the Macroscopic world When the latter is subjected to my Prolificacy Principle"

(see later---Extracts From The Land Of Plenty) ?

Problem numbered one is how to quantize within the Bohr-Sommerfield Rules for the Macroscopic world then how to define the Regional constant? UNIT RADIUS VECTOR (R)

$R = H\xi /k$ Where k=Propagation vector.

Lorentz trans{formation:

$$\Lambda = \begin{Bmatrix} -\beta\gamma & 0 & 0 \\ 0 & 0 & 0 \\ 0 & 0 & 1 \end{Bmatrix}$$

Taking the Lorentz transformation of the four-wavevector is one way to derive the relativistic Doppler effect. The Lorentz matrix is defined as In the situation where light is being emitted by a fast moving source and one would like to know the frequency of light detected in an earth (lab) frame, we would apply the Lorentz transformation as follows. Note that the source is in a frame Ss and earth is in the observing framer, Sobs. Applying the Lorentz transformation to the wave vector:

$Ks = \Lambda^\mu{}_\nu K^\nu{}_{obs}$

Exercise- One:

As always in my ten books R=Unit Vector Radius.

Ω, ω = Frequencies ! S =Source, Obs ==Observer

Translate the above in to the Frequ mechanic System of Physics.?

Answer: $Rs = \Omega_c \, {}^g R \, {}^g{}_{obs}$

Exercise- Two:

Proceeding with the translation obtained above TRANSFORM the following in to the Frequmechanic Systemof Physics ?

and choosing just to look at the $\mu = 0$ component results in

$$k_s^0 = \Lambda_0^0 k_{obs}^0 + \Lambda_1^0 k_{obs}^1 + \Lambda_2^0 k_{obs}^2 + \Lambda_3^0 k_{obs}^3$$

$$\frac{\omega_s}{c} = \gamma \frac{\omega_{obs}}{c} - \beta\gamma k_{obs}^1$$

$$= \gamma \frac{\omega_{obs}}{c} - \beta\gamma \frac{\omega_{obs}}{c} \cos\theta.$$

$$\frac{\omega_{obs}}{\omega_s} = \frac{1}{\gamma(1 - \beta\cos\theta)}$$

where cos θ is the direction cosine of k^1 with respect to k^0, $k^1 = k^0 \cos\theta$.

Source moving away (redshift)

As an example, to apply this to a situation where the source is moving directly away from the observer ($\theta = \pi$), this becomes:

$$\frac{\omega_{obs}}{\omega_s} = \frac{1}{\gamma(1+\beta)} = \frac{\sqrt{1-\beta^2}}{1+\beta} = \frac{\sqrt{(1+\beta)(1-\beta)}}{1+\beta} = \frac{\sqrt{1-\beta}}{\sqrt{1+\beta}}$$

Source movi towards (blueshift)

To apply this to a situation where the source is moving straight towards the ($\theta = 0$), Become

$$\frac{\omega_{obs}}{\omega_s} = \frac{1}{\gamma(1-\beta)} = \frac{\sqrt{1-\beta^2}}{1-\beta} = \frac{\sqrt{(1+\beta)(1-\beta)}}{1-\beta} = \frac{\sqrt{1+\beta}}{\sqrt{1-\beta}}$$

Source moving tangentially (transverse Doppler effect)

To apply this to a situation where the source is moving transersely with respect to the observer ($\theta = \pi/2$), this

The Apple And The Blackhole.

Regional Constants $(H)\mu,\nu$

The General Argument Proceeding on the assumption that Frequency can be signature for energy as evident from planck constant for photons $E=h.\omega$

Then why not for all other objects microscopic or macroscopic if the right constant a regional constant can be predetermined ? But this raise another question of how to define the the (REGION)?Let both the region and its constant identify themselves by themselves Based on the facts that although (Alpha $(\tilde{\alpha})$= 10-(Minus Betta)

The range of $\beta=0$ ∞Thus the region will be a MAP from $(\alpha:\beta)$

$\alpha = (1\quad 0)$ To Betta $(0\quad \infty)$Such that: TRY 3 Dim GREEN THEOREM

F   Omega R=Region

$$(\alpha)\mu = (H)\mu,V.\beta\nu$$

whether the cosmic flux physically exist or not all analysis in my

ΦΦFrequmechanics are group (g)theoric valid and useful:
$V \rightarrow V^*$ Vector Space $\rightarrow$ Covectors $\tilde{\omega}c \rightarrow J.\tilde{\omega}g$

$$\left(\omega \wedge \omega \text{ Gluing}\right) / \left(\omega \wedge \omega\right)_{\text{Cutting}} = \rbrack I \lbrack \Phi \quad (\text{The Identity}).$$

My Theory Having The Potential Of Replacing Newtonian Mechanics.

Isam .T . Saleh. © **September -2016-London.Part-One.**

For Further Details And Backgrounds See My Seven Books Frequmechanics + Frequmechanics Two + Frequmechanics Three + Frequmechanics Four +The Five Arithmetic-S + Fasten Your Seat Belts ?All

published by: htp//kdp.amazon.com

Two Cases the _Classical_ and _Quantum_ Case of Freeparticle climbing upward against gravity from(A) TO(B) Then Falling from (C) to(D) As in the following diagtram :

NOT All are in the plane of the paper.

B $(r.\dot\omega c = r.\dot\omega g \quad) \quad$ C

C   Gluing

. A   → Direction of Flux   D

$\dot\omega g >> \dot\omega c$ $\dot\omega c << \dot\omega g$

The Kinetics Of Frequmechanics:

My equation In its basic form:

$$\Omega = (\omega_{cutting}) \pm (J\omega_{Gluing}) = \Delta\omega$$

Let us consider quantum particle e.g. photons or **Emitons**© The infinitesimal $\Delta\omega \rightarrow 0$ Giving rise to the followings: $\pm J = \omega_c / \omega_g$ But we already defined the lag or phase angle > $\Phi_{Emitons} = \omega_g / \omega_c$ Therfore $\Phi = 1/\pm J = \pm J$ eqn-A

But (J) In itself can only become physically meaningful positive ifraised to the powers of (4n) Where n=1,2,3,4...etc).

Giving that : $\Phi \rightarrow \Phi = 1$ at all $(\Phi)^{4.n}$ As in the diagram below :

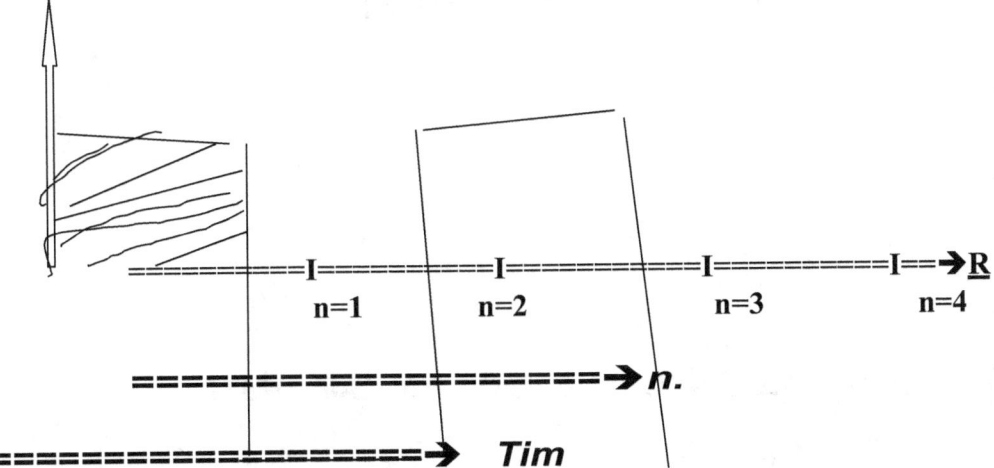

To clarify and hopefully justify the frequmechanic case we first present the **Quantum Picture** !Followed by the Classical picture!! Then the frequmechanic Picture .

The Quantum Picture.

Separating of the wave equation: For Free Particle:
The wave equation (Schrödinger -3D).

$$ih \cdot \Psi(r,t) = -h^2/2m \nabla^2 \cdot \Psi(r,t)$$

Can be rewritten if the potential V (r,t) does not depend on momentum.

$$ih \cdot \Psi(r,t) = -h^2/2m \nabla^2 \cdot \Psi(r,t) + V(r,t).$$

$$\Psi(r,t) \rightarrow U(r) \cdot f(t)$$

$$i.h/f \ (df/dt) = -1/U \ [-h^2/2m \nabla^2 \cdot U + V(r).U\,]$$

$$f(t) = C \cdot e^{i.E/h}$$

$$E \ U(r) = [-h^2/2m \nabla^2 \cdot + V(r).] \cdot U(r,)$$

Boundary conditions at infinite potential where:

$$\Psi(r,t) \rightarrow U(r) \cdot e^{i.E/h}$$

$$ihd\Psi/dt = E \cdot \Psi.$$

Boundary conditions at great distances:

$$-h^2/2m \cdot d^2u/dx^2 + V(x) u = E u$$

At $x < 0$ $V(x) = 0$ At $X > 0$ $V(x) = V_0$ If $V_0 \to +\infty$ and
$0 < E < V_0$ Then the Solution :

$A \sin \bar{Y} x + B \cos \bar{Y} x$

$X < 0$ $\bar{Y}^2 = [2mE/h^2]$

$U(x) = C e^{-\beta x} + D e^{\beta x}$

$x > 0$ $\beta^2 = [2m(V_0 - E)/h^2]$

For E<0 the sines and cosines can be replaced by Hypoboralic Sines and Cosines for the solution at x= 0. This consistent ith solutions since its in the vicinity x=0

Thus the particle hit rigid s inpenetralable durface analogous to classical particle with FINITE ENERGY Reverses its momentum on hitting this surface

Referenig back to Eherfest theory and the definition f0r Position probability density and prob current density Pr and Sr the contiuity conditions of P®=Probability density and S®Probabikity current density we have: P(x) And S(x) Vanishes at x=0

Therefore du/dx not zero even at x= 0 compare dh/dt = g.

The classical Picture :

One dimensional classical analogue to the above was confirmed by recent experiments on microscopic particles suspended by complex fields eventually falls in straight line due to gravity therefore its reasonable to argue as follows :

ANALOGUE

Classical (macroscopiv free partice in gravitational field $=E=mgh$ h for height#

Let $r=C.t.$ C=Speed of light.

$$E. U_{(r)} = [-h^2/2m \nabla^2 + V(r)]. U(r,)$$

mg→ m.d2/dr2→1/C2 .d2/dt2 .Ur

based on the reasonablw assumption that all particle Microscopic Emitons (Particle of Frequmechabics.)

Or macroscopic path under gravity can be vapproximated to be straight such that

Let :

V0 =The minimum potential of V(r) Where the energy E>V0 and h runs from +h to -h,

Now thre is one fundamental relationships sgare by both quantum particle and classical(Macroscopic) particle
When $U(+x) = U(-x)$ then the slopes are not equal and whe
The slopes and subsequently the accelerations will be equal
$U(+h) \neq U(-H)$ The slopes are $dmg.h/dt$ $du/dx \rightarrow 1/c$
(du/dt Keeping in mind that the expectation value depends only on time.

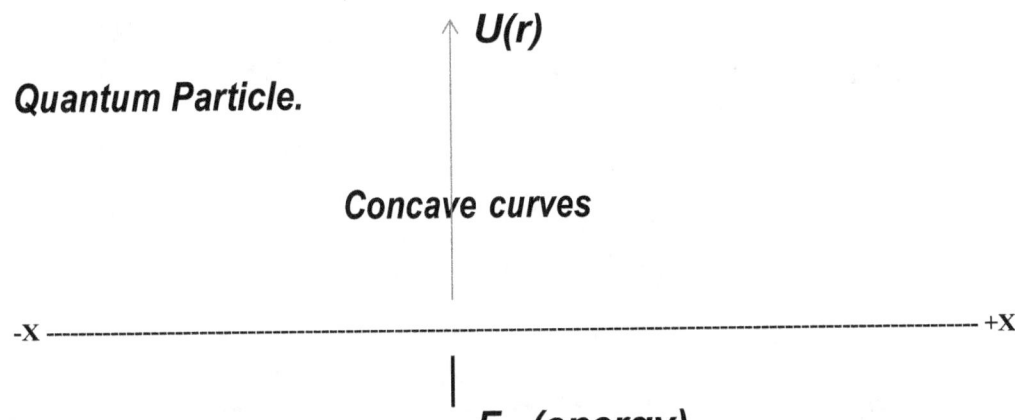

Quantum Particle.

$U(r)$

Concave curves

-X -- +X

--- E (energy)

Courtesy: Q.M.- L. I. Schiff-- 1956- Page:35,36.

$|V_0$=Minimum (Potential of V(r)

Classical Particle   (mg.h)

Concave.

-C.X -- +c.X

Frequmechanic: Emitons.

$m (r. \Omega)^2$

CONVEX

C.J/ ωg -- C/ωC

Courtesy: Q.M.- L. I . Schiff-- 1956- Page 35,36.

Recall ω=Frequency and omega tilde $\tilde{\omega}$=for Angular velocity C.X are values not directions:

$j.r \rightarrow \omega j.r = \tilde{\omega}g$ AND: $\rightarrow \omega.r = \tilde{\omega}c$

That is why we stated inmy frequmechanic books 1-5 that: $\Omega = \omega c + J.\omega g$

And only after multiplying through by (r) we have:
$$\Omega^{Tilda} = \tilde{\omega}_{Cutting} - \tilde{\omega}_{gluing}.$$

Meanig at the tip top of the trajectory : $\Omega^{Tilda} = 0$ Because :

$\tilde{\omega}_{Cutting} = \tilde{\omega}_{gluing}.$

While the acceleration $g = d2h/dt2$ remains constant as shown in the diagram above that when the eigen function $Uc(+X) \neq Ug(-x)$ the slopes and subsequently the centripetal accelerations will be equal and perpendiculat to the angular velocities justifying rRELATIONSHIP TO THE EIGEN FUNCTION (U)Described bybthe frequencies as shown in the diagrams when e by definition

$$\omega c \neq \omega g$$

Except at the tip top of the trajectory where the accelarations change directions .

Keeping in mind C.t and C/ωC or J,C/ωg

Are values not directions and that C/ωC or C.j/ωg represented by convex not concave curves yet the same argument applys and we are deaing with particle trapped in potentialVr unable to escape to infinity (First class Wavw).

Exercise from The AboveGraph ?
If the slope dU/dx are wquals at two points on two different curve does that iimply
Classical particle =Quantum particle ?

****

The Frequmechanic Picture ;

In contrast to the treatment of my theory in my other four books (Frequmechanics One, Two, Four, And Five) Now in this new edition of Frewumechainc Three as the name imply we express my theory in 3- Dimensional analysis by considering The cross Product:

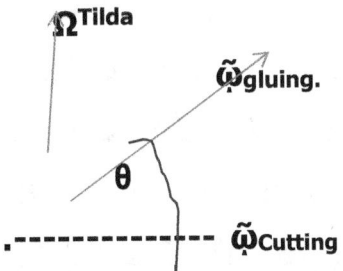

At the bottom trajectory (The starting point of free particle) $\theta = \pi t$
the tip top of the trajectory $\theta = 0$ thus

$$\Omega^{Tilda} = \tilde{\omega}_{Cutting} \times \tilde{\omega}_{gluing} \cdot \sin\theta$$

The centripetal acceleration :

$a = dv/dt = d(\Omega^{Tilda})^2 \cdot r/dt = 2(\Omega^{Tilda})d(\Omega^{Tilda})/dt$

$a = \tilde{\omega}_c (d\tilde{\omega}_g)/dt + \tilde{\omega}_g (d\tilde{\omega}_c)/dt] \sin\theta + \tilde{\omega}_{Cutting} \cdot \tilde{\omega}_{gluing}) (-\cos\theta) d\theta/dt$

$a \cdot dt = \tilde{\omega}_c (d\tilde{\omega}_g) + \tilde{\omega}_g (d\tilde{\omega}_c) \sin\theta + (\tilde{\omega}_{Cutting} \cdot \tilde{\omega}_{gluing}) \times (-\cos\theta) d\theta$

Integrating coth sides :

$a \cdot t + \text{Constant} = \tilde{\omega}_c (\tilde{\omega}_g) + \tilde{\omega}_g (\tilde{\omega}_c) \sin\theta + \tilde{\omega}_{Cutting} \cdot \tilde{\omega}_{gluing}) \int \times (-\cos\theta) d\theta$

$a \cdot t + \text{Constant} = 2(\tilde{\omega}_c \tilde{\omega}_g) \sin\theta + \tilde{\omega}_{Cutting} \cdot \tilde{\omega}_{gluing}) \times (-\sin\theta)$

$a \cdot t + \text{Constant} = (\tilde{\omega}_c \tilde{\omega}_g) (2\sin\theta - \sin\theta) = (\tilde{\omega}_c \tilde{\omega}_g) (\sin\theta) = \Omega^{Til}$

Therefore $a = \Omega^{Til}/t - \text{Cons}/t$. wrong

Now at the beginning of the pathe for free particle
i.e at θ=0 Sin θ =0 a=Constant i.e the constant
(g)Garavityacceleration. And ar
θ= π/2 at the top of the trajectory Sin π =0
Again implying that a=g

So where is the break through?We provided an expression for all kinds of potential by frequencies only (No distances or time etc). As we said earlier in this book (Frequmechanic three)The particle assumed to fall approximately in straight line inside gravitational field Therefore establishing that

θ=0 OR π/2 hence (g)

But not all potential field provide straight projection.g (electric electrostatics etc)Can operate on particle S at different (θ)>Again only frequencies are required to determine the potential effects.equipotential or not.

$a = (\tilde{\omega}_{Cutting} - \tilde{\omega}_{gluing})/t$ -Constant $a = \delta\tilde{\omega}$. – Constant

    Clearly this constant =g (Gravity Acceleration)when $\delta\tilde{\omega}$ 0 ·

    But why should $(\delta\tilde{\omega})$ Not Vanishes→0

Macriscopically ω_c and ω_g are very close ly tailing each other

2-By definition the unit raddie stretching fro2 m infunitum to each
    frequencies $R_c \approx R_g$ Therefore $\tilde{\omega}_{Cutting} \approx \tilde{\omega}_{gluing}$.

IMPORTANT RESULT:

The term s for $\delta\tilde{\omega}$. becomes very important when we move from the MACROSCOPIC to the MICROSCOPic:

TRY ENRGY ROUTE :

$\Omega^{Tilda} = \tilde{\omega}_{Cutting} \times \tilde{\omega}_{gluing}. \sin\theta$

$\Omega'.R = \omega c .rc \times J\, \omega g.rg \sin\theta$

$m.(\Omega'.R)^2 = m(\omega c .rc \times J. \omega g.rg \sin\theta)^2$

$.(\Omega'.R)^2 =- (\omega c .rc \times \omega g.rg)^2 + ((\omega c .rc \times \omega g.rg)^2 (\cos\theta)^2$

$(\Omega'.R)^2 + ((\omega c .rc \times \omega g.rg)^2 (\cos\theta)^2 - 1)$

$\Omega'.R)^2 = - ((\omega c .rc \times \omega g.rg)^2 (\sin)^2$

Giving : $\Omega'.R)^2 / - ((\omega c .rc \times \omega g.rg)^2 = (\sin)^2$

But: $.(\Omega'.R)^2 =- (\omega c .rc \times \omega g.rg)^2 + ((\omega c .rc \times \omega g.rg)^2 (\cos\theta)$

$\cos\theta)^2 ==: \{(\Omega'.R)^2 + (\omega c .rc \times \omega g.rg)^2 \}/ (\omega c .rc \times \omega g.rg)^2 +$

$\tan\theta)^2 = \Omega'.R)^2 / - ((\omega c .rc \times \omega g.rg)^2 \} // \{(\Omega'.R)^2 + (\omega c .rc \times \omega.rg)^2 \}/ (\omega c .rc \times \omega g.rg)^2$

But $\omega g/ \omega c = \tan\theta$

EXERCISE:

Which of the twopossible values : Θ (0 V π/4) While $\Phi =$(0 V 1) are considered by Frequmechanic the symbol (V) for (OR) Substituting for $\tan\Theta$ (0 V π/4) Φ(0-1)

VERY Important RESULT.

One very important result from the previous section (THE LIMITS OF PHASE ANGLE) That :

No matter how close the two frequencies (ωc & ωg) Even at the point when both are identical in ‖Value‖ The Phase angle $\Phi = \{(\omega g / \omega c)\} \neq 1$

Implying there is : $\Phi - \delta\Phi = (\omega g - \delta \omega g)/\omega c$

: $\Phi \cdot \omega c - \omega c \cdot \delta \Phi = (\omega g - \delta \omega g)$

$(\omega g / \omega c) \cdot \omega c - \omega c \cdot \delta \Phi = (\omega g - \delta \omega g)$

$(\omega g / \omega c) \cdot \omega c - \omega g = \omega c \cdot \delta \Phi - \delta \omega g)$

$0 = \omega c \cdot \delta \Phi - \delta \omega g)$

$\omega c \cdot \delta \Phi = \delta \omega g)$

In the li mits of (1,0) fter integration:

$\omega c \cdot d\Phi = d \omega g) \rightarrow$

$\Phi = \omega g / \omega c + K(Constant)$

Meaning EVEN when :

$$\omega g = \omega c \ ?$$

Φ Remains Constant ≠ 0.

Which proves my point that EVEN at Equilibrium Φ and by implication (g) Do not vanish.

Although in theory Φ Max =1.

Yet by definition Φ<1. (<u>Infinitesimally</u> less than one)

Hence the name (Lag Angle) Often used instead of Phase Angle. Meaning NATURE WILL NEVER ALLOW Φ=1 Just as C=Constant speed of loight. Thus oue Knowledge of the universe destinrd to have INTRINSIC LIMITATIONS.

Clearly there is DEGENERATIVE Process in the energy levels of cutting and gluing the C.F. And that its <u>FUNDEMENTAL RULE OF NATIRE</u> therefore at this point we need to turn to Q.M.

Quantum Mechanic.

Now we already established in the previous section that the range of Tan Θ is that of Sin Θ Which is :

$$(0 - \sin \pi/4.)$$

While still on the topic of 3D(r) and the orthonormality of the two components of Ω^{Tilda} Its possible to extend the arguments to any other eigen function U0 by choosing the corresponding Coefficient's (and b) Such that :

$$U = a.U_c + bU_g .$$

.a./b = $\int [u*g(r).u_c(r) d3 /(u_c)^2] d3r$

$\int *c(r).u_g(r) d3 /(u_1)^2 d3r = \delta_{c,g}$

If we indicate the degeneration by (n) For Energy (E) Then:

$$\int U*E_g (r) u E_c (r) d3r = \delta_{E_c, E_g}$$

- $\int U*E_g,n_g (r) u E_c,n_c (r) d3r = \delta_{E_c, E_g} . \delta_{n_c, n_g}$

The Proof:

Three of my Equations in BASIC Forms
Foe Frequencies : $\Omega = \omega_{cutting} \pm J\omega_{Gluing}$

$\omega c = \omega\text{Cutting} \quad \omega g = \omega\text{Gluing}$

For Angular velocities: $\Omega^{Tilda}\text{Cutting}) - J.(\tilde{\omega}_{gluing})$

And in 3D: $\Omega^{Tild} = \{\tilde{\omega}_{Cutting} \times \tilde{\omega}_{gluing}\}\sin\theta.$

A note about the indices I am about to use in this book are of different functions and meaning to the same indices used in Frequmechanics -Five. Also due to computer/typing Restrictions special care need to be given for example (uv)Appears to be on top of each other when both (Mu and Nu)Should be together etc?etc? Sometimes written μ (Mu)=u and v(Nu-v)

Let: $\Omega^{Tilda} = \Omega^{\mu,v}$

$\tilde{\omega}_{Cutting} = (\tilde{\omega})^{\mu}$ AND $\tilde{\omega}_{gluing} = (\tilde{\omega})^{v}$

$\Omega^{u,v} = (\tilde{\omega})^u \pm J.(\tilde{\omega})^v$

Differentiting Ωuv w.r.t   EACH : ωu And ωv respectively then equating the results

Let :

$$\mu - N = m \quad \text{And} \quad n = v - N$$

Where (N) Is the number of times Differentiated :

$$d^N \Omega uv / d\omega u = u \cdot \omega^m \pm 0 \quad \text{-------------------- EQN-A}$$

$$d^N \Omega uv / d\omega v = 0 \pm j \cdot v \cdot \omega^n \quad \text{------------------- EQN-B.}$$

Now the two angular velocities ($\tilde{\omega}^\mu$ and $\tilde{\omega}^v$) have two functions:

1-

By definition they must remain different at all time i.e
$\mu \neq v$

2-

By the Physics of frequencies, they must remain <u>Infinitesimally</u> close to each other ?

(Subtle point is frequencies not measured in fraction of whole numbers yet they must be close not between (0-1)) i.e

$\mu \approx v$

Thus; $\quad d/d(\tilde{\omega})\quad \overset{u\,=}{d/d(\tilde{\omega})v}$

Giving:

$$d_N(\tilde{\omega}_g)^v / d_N\,\tilde{\omega}_c) \quad \overset{u\,=}{\pm (u/J.v)N\,\{}\overset{m\,/}{(\tilde{\omega})}\overset{n\}}{(\tilde{\omega})}$$

Therefore:
$$\Phi = \pm u/J.v\,^{m-n.}(\tilde{\omega})$$

Now for brevity if we had RE-Differentiate the above equations :

N-times we have:

$$\Phi^N = \pm (u/J.v)^N\,^{m-n.}(\tilde{\omega})$$

Since m= μ-N and n=v-N

Q.M.

The capital letter (N) Is function of height for free falling classical particle with mass (m) in gravitational field (g) Such that : EXPECTATION VALUE =

$$\rightarrow$$
$$< \tilde{\omega}_g{}^{**} \mid d_N/d\Omega \mid \tilde{\omega}_c >$$

Where N= Range from 1-h
At the top end of the trajectory when the particle come to rest
$\tilde{\omega}_{gluing} \rightarrow \tilde{\omega}_{cutting}$

$$\Phi - (\delta)^N = \tilde{\omega}_g / \tilde{\omega}_c$$

Taking Logarithms :

$$\text{Log} \{\Phi - (\tilde{\omega}_g / \tilde{\omega}_c)\} = N \cdot \text{Log}(\delta)$$

Substituting for:
$\Phi = 1$ at $\tilde{\omega}_c = \tilde{\omega}_g$)
N= {Log (1-Φ)}/ Log(δ
N= Log {(1/$\tilde{\omega}_c$) .Φ) Log δ
N= Log (1- δ – Φ)
Now substituting for N= || h || h= (Height)
Where Energy E = m.g.h thus
h= E/mg

$N = \|h\| = \text{Log}(1 - \delta - \Phi)$

Thus my Energy Equation Of Motin in terms of the phase angle:

$E/mg = \text{Log}(1 - \delta - \Phi)$
At $\omega c = \omega g$ $\Phi = 1$ Therfore :

$E/mg = \text{Log}(-\delta)$ → $E = mg \cdot \text{Log}(\delta)$

Thw Embedded (δ)

1-For any frequency Ω The observer measure it from their own frame of reference as F (observer) While the source sending it at F(source) and the difference : $Fo - Fs = \delta$ And the probability of measuring (δ) Will be same as that of energy as we shall see later?
Now one of the principles constructed by Frequmechanic is that Frequencies are always measured in full numbers and its Meaningless to state frequency as (0.5)Cycle per second or as HALF Herz.
 Therefore the final total count of frequencies always rounded up by adding or shrinking (Absorbing) (δ). Thus the (Fraction).Are embedded ?Its when Physics not always obey mathematics.
Thus δ re present that embedded <u>Physical</u> Fraction !
Resurfaced by Frequmechanic to make use of it for further study?

And this basically one the Physics of Frequmechanics.
   This is not definition! Its due to the measurement of frequencies.
b-Norice that any term containing $(\delta)^N$ does not vanish except:
Norice that any term containing $(\delta)^N$ does not vanish except:
Wnen N→∞ *such as nearer to blackholes*
When N→∞

__RESUL:__

Clearly EVEN if m=n i.e u=v or

$$\tilde{\omega}_{Cutting} = \tilde{\omega}_{gluing}$$

The Phase Angle (Φ) Not allowed to reach ONE.

And can oly move to POSITIVE terotory whe raised to the Furth power.as shown in the following diagram ?

For deeper meaning of these resultee the bottom line at the end of this b

Now this is the answer why

Φ not allowed to touch ONE.

Since if Φ=1 (Exactly) It means that e µ= v

And it followas that

$(\tilde{\omega}_{cutting} = (\tilde{\omega}_{gluing}$ implying there will be no Cosmic Flux ti Cut or to Rejoin which dey the object of the exercise!

As $\tilde{\omega}_c \rightarrow \tilde{\omega}_g$ Φ→1

and at $\tilde{\omega}_c \neq \tilde{\omega}_g$ Φ <1.

Q.M.

"The **PURE STATE** represented by $\Phi = 1$ at $\tilde{\omega}_c = \tilde{\omega}_g$!"

The capital letter (N) Is function of height (h) for free falling classical particle with mass (m) in gravitational field (g) Such that:

EXPECTATION VALUE where the omega tilda-s NOW velocities of UNIT mass:

= $<\omega_g {}^{**} |\, id/d\Omega\, |\, \omega_c>$

Where N= Discrete Number Range from 1-h

At the top end of the trajectory when the particle come to rest therefore:

$\omega_g \rightarrow \omega_c$

Introducing The (δ) Switch:

such that: Phase Angle = $\Phi \cdot (\delta)^N$

$\Phi = \omega_g / \omega_c$

$(\delta)^N = 1$ At : $\omega_g = \omega_c$

$(\delta)^N = (0-1)$ At $\omega_g \neq \omega_c$

$(\delta)^N = 0$ As $N \rightarrow \infty$

First Route:

Phase Angle ≈ 1

Phase Angle = $\Phi - (\delta)^N$

$\Phi - (\delta)^N = \omega_g / \omega_c$

Taking Logarithms:

$\text{Log}[\Phi - (\delta)^N] = \text{Log } \omega_g - \text{Log } \omega_c$

ALSO:

$\omega_c - \omega_g = (\delta)^N$

Log ($\omega_c - \omega_g$) = N Log (δ)   Giving:

N= Log ($\omega_c - \omega_g - \delta$) Therefore at $\omega_c = \omega_g$

$\qquad$ N = Log δ

We already said that h(Height)=F (N)'

Energy =mg.h = m.g.Log δ Giving:

Energy /mg = F(Log δ.)=H. Logδ

Where (H) Some universal Constat cpnfirming my search in all my other books for this Universal Constant.

For Energy E: $<E'> = <\Phi^* | i.H.d/dt | \Phi>$ And for Energy

E': $<E'> = <\delta^* | i.H.d/dt | \delta>$

$$[-ih/2m \nabla^2 -(V(R))] U®E = E\, U(R) \text{---- EQN-A}$$

$$[-ih/2m \nabla^2 -(V(R))] \delta^* RE' = E' \delta^* (R) \text{ ---EQN-B}$$

First Linear Route Subtracting A & B.

This in 3D After some manipulation eliminating potential (V) And multiplying by multiplying EQN-A with δ* and EQN B with u Eigen Functions.

$E - E' = -ih/2m \int (\delta^* \nabla^2. U - U. \nabla^2 . \delta^*)\, d3r / \int \delta(r) . \delta(r) .d3r$

All we done so far is translating the linear expression ($\Phi - \delta$) By applying Q.M. in to The difference between two energy ($E - E'$).

Norice that any term containing $(\delta)^N$ does not vanish

Second Route Multipying Equations A & B.

This is left as Exercise proceeding from: Phase Angle = $\Phi \times (\delta)^N$

Instead of Phase Angle = Φ -Minus $(\delta)^N$ Keeping in mind that in this case the potential (V)only Vanishes when (δ)=0.

Clearly The term $(-1/J.v)^N$ can only become positive when (N) is discrete number at N>4 as shown in the GRAPH below. :

N=+4,+16, etc ?

Master Graph Of Frequmechanic.

$F(\Omega)$
$Y(n, J\underline{\tilde{\omega}g})$

$$d/dt < \Omega^{Tilda} > = (1-J) \, d/dt<\underline{\tilde{\omega}}>$$

Discrete Quantum Potential.
(Positive Field)

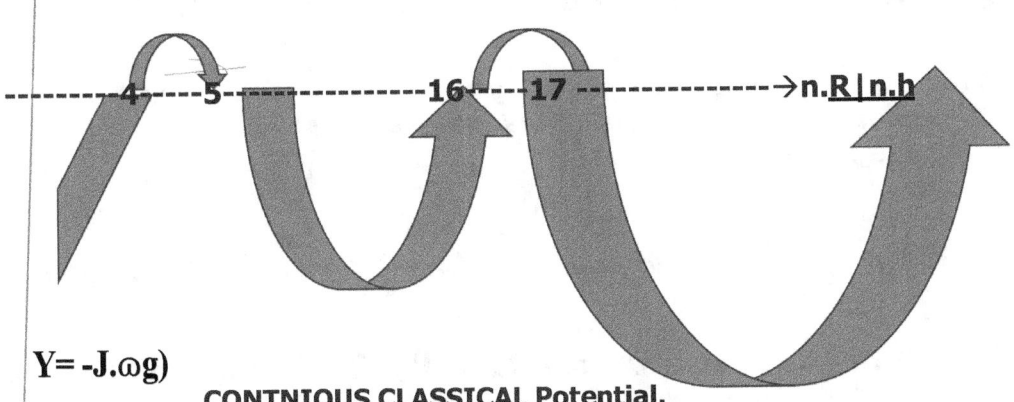

----4----5--------16----17-------------→n.$\underline{R}$|n.h

$Y= -J.\omega g)$

CONTNIOUS CLASSICAL Potential.

(Negative Field)   →→ X (N.$\underline{\tilde{\omega}c}$)

$Y= \delta F(\Omega c,g.)$ X= N.h =Multiplate of (h-Bar)Planck constant. OR:
$Y=n.\underline{R}$ =Multiplate of radius vector R .AS defined in all my books?
n=1,2,3,etc?

Evaluation Near Infinity.

1- The above curve originating from the Negative complex imaginary lower Field to surface at n=1 (4^1)Rising to the Real positive field then Ducks at 4+1 =5 Back in to the Negative Field to surface again at n=2 4^2 =16 then DIPS again in to the Negative Field at 16+1=17 only to surface again at (4^4) AND SO FORTH to infinitum.

2- All we know about this curve is by definition must be smooth Differentiable but we do not know its shape or the areas it subtend above or below the X-axis.

2- The y-axis =Frequencies $\omega_{cutting}$ and ω_{gluing} R=Hypothetical radius vector from infinitum. as defined in all Frequmechanics books.

3-

Angular velocities per any wave packet= $t1 \sum^{t2} \omega c.n.R - J_{x1}\sum^{x2} \omega g. n.R$

Energies PER wave PACKET = $t1\sum^{t2} n.h. \omega c - j._{x1}\sum^{x2} n.h. \omega g.$

Since the **CONNECTING** curve by definition must be smooth Differentiable therefore Total areas representing :
Total Area PER ANY PACKET. $A = \int y.dx = \int F(\Omega).dx$.

$A = t1\int^{t2} \{ t1\sum^{t2} \omega c.n.R \}dx. - j_{x1}\int^{x2} \{_{x1}\sum^{x2} n.h. \omega g.\}dx.$

Limits of the first integral: $t1= 4^n$, $t2=4^n+1$. Limits of the second integral: $x1= 1+(4)^n$, $x2= 4^{(n+1)}$ Notice $t2 = x1$ i.e Continuity? Now let us approximate this mysterious curve with circles:

The upper positive quadrant consist of HALF circles of equal areas. Diameter =In Unit Length Radius (U).

Where: $U^2 = 1$ → Total are $= N.\pi.1/8$.

The lower Negative quadrant the curve describe unequal HALFcircles:

Diameter $= (x_2-x_1)$ Or $= 4^{(n+1)} - \{1+(4)^n\} = 4^{(n+1)} - (4)^n - 1$

Area $= \pi.(1/2).U \{ \sum_{x_1}^{x_2} U.(4^{(n+1)} - (4)^n - 1)/2 \}^2$

Total Areas described by this curve: $u^2 = 1$

$= (\pi/2) [(N/4.u^2 + \{ . {_4\sum_N} \ U.(4^{(n+1)} - 4^n - 1)/2 \}^2$

Now let $(\pi.Z^2)$ Be equivalent area to the above total:

$\pi. u^2 Z^2 = (\pi/2). [(N/4. u^2) + \{ {_4\sum_N} \ U.(4^{(n+1)} - 4^n - 1)/2 \}^2$

Eliminating (π) and (U) Now: Z in dimensions of <u>Units Of Length</u>:

$Z = [(N/8) + (½)\{ {_4\sum_N} \ (4^{(n+1)} - 4^n - 1)/2 \}^2]^{1/2}$

Z Now can be SCALED to give meaningful finite physical values against those in the lower quadrant that blow up extremely large on its way to infinity. Thus we established MECHANISM for <u>TRACING</u> large values.

<u>EXERCISE:</u>

If Diameter Length=one unit of (X) And $Z^2 = X^2 + Y^2$
(R)= X=Y=1/2 Giving $Z = 1/\sqrt{2}$ _Evaluate (+x) At Y= - 63/2 ?
Does the <u>RATIO</u> of total areas in the upper to the lower fields up to any point on the X-Axis remains constant? And How ?{1/m.R}and (m=?).

Solution.

Small n=1,2,3,etc.
Number of upper circles =Capital N
Number of lower circles= N-1.
Radius of the equal uppe$(4^{(n+1)} - 4^n - 1)r$ circles =r.
Radii (Plr) Rn =multiples of r =n.r.
.m= ωc - ωg (Frequencies not Tilda Angular velocities).
Then the <u>Ratio</u> of upper area to the lower at any point on the X-Axis
= $\pi.r^2 / [\pi.m.(n.r)^2.] = 1/m.n^2$
Therefore, as n→∞ AND   m→0 When ωg →ωc
Hence the <u>Ratio Remais constant</u>
$d/dt <\Omega^{Tilda}> = (1 \pm J) \, d/dt<\tilde{\omega}>$

$d<\Omega^{Tilda}>/d<\tilde{\omega}> = (1-J)$ constant.

Transferering Discrete To Continuous.

From The Slope Of Master Graph For Frequmechanics:

. $dy/dx = d\Omega^{Tilda}/dR = d\Omega n \, R/dR = \Omega n \, dR/dR = \Omega n = (1 \pm J)\omega$
Ωn =Natural Frequency (Not Angular Velocity (Ω^{Tilda}) Therefore:
$$.dy = =(1+J)\omega . dx.$$
Integrating:

$Y= =(1 \pm J)\omega.x + \Phi$ Giving :
$$\Phi = [<y> \pm \omega <x> \pm J.\omega <x>$$

<u>This Is Important Result Since We Can Predict Values Near Infinity And Vice Versa. i.e. We Now Have Scale To RETRIEVE informations lost at NEAR INFINITUM.</u>

The Philosophy.

"Since the probability of finding any particle in infite space is one therfore we ask : Can there be any correlation between spatial stations and those of probabilities by back tracking any stochastic process ?

The answer to the above question is NO! Since it contradicts the very definition of probability and erase the concept of pure state in Q.M.
However in classical probability we can predict that the probability of person taking ill is function of the amount of alcohol or number of cigarettes taken.
The above argument raises one question that of the continuum of classical in to quantum mechanics
in this section we deal with that question it can be addressed by moving to three dimensional analysis i.e moving from that of areas to spheres
the answer is to find bridge from classical to the quantum:
This was already given in my theory (<u>The Prolificacy Principle</u>)under the title :
(Land of the Plenty) in my other books on Frequmechanics :
Briefly quantum effects can also occur in the Macroscopic world if there was enough numbers of exactly identical Macroscopic copies of particles i.e the problem is shifted to that of numbers .
Instead of finding the probability of of finding particle in a box we now go to the probability of finding particle in sphere and instead of letting the box determine the Probability we say the sphere will contain the issue! Meaning the radius of the sphere and the probability are related by $r=1.=Prob.$!?!?!

The Science.

"Here we are dealing with abelians multiplications of unite i.e ab=ba therefore Caution required not to confuse :
n with N. OR the Frequency Ω^{Tilda} with Four vector velocity $\Omega^{SPHERICAL}$

.n= Number of volume units Also

.n=Number <u>Microscopic</u> of observables

.n =the position along the X-Axis =Time Axis.

N=Periodic vibrations of Frequencies Also

N=Number of cycles describibng each sphere. Also

N=Number of <u>Macroscopic</u> observables.

Clearly both dimensions and units are mixed unorthodoxly but this soon will be overcome once we get familiar with interpreting the above MASTER GRAPH OF FREQUMECHANICS in multiple roles for different situations.there numerous examples of that but here there is only spec for three:

1-
If Y-axis represent any number of cycle of any periodic movement and the X-Axis now represents units of time then:

The Slope dy/dx =Φ The Phase Angle.

2-
If (n) on the X-axis can be taken as the number of unit volume (v). While the Y-Axis can represent the number of Microscopic Particles in (V) Slope of the graph =the expectation value for such particles.

3-
Y-axis can represent the expectation values while the X-Axis Represent (n)=Number of (v)Insise macroscopic Particles :Slope=Probability Density Current.
(Search The Prolicicacy Principle Where :
$$N/Z_{(A)} = K. = 1/500 \times 10^{+27}$$

$N = ((4^{(n+1)} - 4^n - 1)^2 / 4$

$= \frac{1}{4} [4 - 1 - 4^n) 4^n]^2$

$N = \frac{1}{4} [3 - 4^n) 4^n]^2 \rightarrow$

$4^{2.n} / 4 [9 - 6. 4^n + 4^{2n}]$

At n=0
$N = \frac{1}{4} (4) = 4/4. = 1$

At n= 1 $N = 4.(+ 1) = 4$

At n= 2 N

Spherical Ensembles And Equipotetials.

in the real physical world matters are made of many particles not one. The above argument raises one question that of the continuum of classical to quantum mechanics this question can be addressed by moving to three dimensional analysis i.e moving from that of areas to spheres and the box is replaced by sphere where the wall of the box replaced bi the sphere cutting each at (X+!)As the periodic boundary conditions !(See above Special Relativity 3D Hypersurface).

THUS GIVING INFINITY A FINITE LABELS though NOT value :

Therefore L3 replaced by r3 etc.as follows:

That is to say the half circles in the upper quadrant now describe full sphere of unit volume (v) ?
$$V = \tfrac{3}{4}\pi\, r^3$$

Let r the radius for the the upper circles =1 CM unit of length.
Volume of theses equal Spheres :
$V^{upper}(V^u) = 3\pi/8 \ CM = v/2$

Total $V^{uppe} = v(n-4) = (n-4)$ Unit Volume.

Total $V_{Lower} = n1.v + n2.v + n3.v$ etc Where $n1, n2, n3$ are now coefficients defined as follows:

Total $V_L = v(n1+n2+n3..) = n1+n2+n3.. = \Sigma n$.
Such that: Now we can measure the lower volumes in multiples of (v)
Total $VLower = V_L = N.\ v$
Total $= N1.v + N2.v + N3.v....$etc. Where $N2.v - Ni.v > 1$. And their expectation values :
$\{<NJ.v> - <Ni.v>\} = <N>v \quad \{<N> - <Ni.>\}.V = <N>v$
$\{<NJ.> - <Ni.>\}/> = <N>$
As Both Ni and Nj becomes too large the probability of their difference of reaching N is one
A number $<NJ.v> - <Ni.v>/<N>$ its same as saying:
Let T=Unit Time: $T<\omega c> - <j.\omega g> = <\Omega>T$

Implying that no matter how huge spheres Ni &Nj or Frequencies ωc & ωg become the differences between them becomes FINITE.
The abpve step requires SUBTLE and CONCEPTUALY DIIFICULT operations as follows:
First recall we are now dealind with sheres i.e:
Rx =Ry=Rz=Rc=Rg=Rtelda=thathypothetical vector radius R=1.
Then multiplying through the above expression by R we obtain expression for angular velocities :

$$< \tilde{\omega}c > - < j.\tilde{\omega}g > = <\Omega^{telda}>$$

The next step how to relate these velocities to the flow of the cosmic flux that been cut and rejoined at the above velocities
{<N > - <Ni.>}.Volume =<N>Volume
The next step is to show <N>V=Finite?
Tis take us to the very definition of expectation value which can be rewritten:Then from the Eqn for the flow of no sourc and no sink.

$$_0\sum^4 \quad _4\sum^\infty \quad \{ \int \nabla \times n.F \, dA = \int_c F.dl \}$$

Let: $L1 =_0\sum^4 \{ \int \nabla \times n.F \, dA \}$

$L2 =_4\sum^\infty \{ \int \nabla \times (n.F \, dA)^2$

Integration round the Curve (C) is REPLACED by Circles cutting © and glued(g) per unit time.giving:

$\int_c F.dl \}.\int_g F.dl \} = L1.L2$.

$.L2 =_0\sum^4 \times_4\sum^\infty \quad \{ \iint \nabla(n,n.F \, dA . \nabla F \, dA\}$

Since At $0 \to \infty$ The expectation <N>
L1.L2 = <N>Volume =Constant of integration.
$= \int_{cutting} F.dl . \int_{gluing} F.dl \}$
Lc-Lg = The difference between the two constans of integration i.e The difference is FINITE .Without the need to know the limits of thee TWO integrations at values NEAR infinity.

$N = ((4^{(n+1)} - 4^n - 1)^2/4 = \frac{1}{4}[4 - 1 - 4^n)4^n]^2$

$N = \frac{1}{4}[3 - 4^n)4^n]^2 \rightarrow 4^{2 \cdot n}/4[9 - 6 \cdot 4^n + 4^{2n}]$

At $n=0$ $N = \frac{1}{4}(4) = 4/4 = 1$

At $n=1$ $N = 4 \cdot (+1) = 4$

At $n=2$ $N=$ etc------

3D Schrodinger Equation:

$\{-\hbar^2 \nabla^2/2m + V(r)\} u(r) = E \cdot U(r)$

Equipotential means : $V(rx) = V(ry) = V(rz)$

Momentum (P) means directions: and the The solarium:

$i\hbar \nabla \cdot U_p(r) = P$ If $C =$ Normalizing Constant:

And $K =$ Propagation vector $:= P/h$

$U_p(r) = C \cdot e^{i(P \cdot r)}/h$

$U_p(r) = r^{4/3} \cdot e^{(iK \cdot r)}$

$U_p(r) = C \cdot e^{(iK \cdot r)}$

Sphere Normalization:

$K_x = 2 \cdot \Pi \cdot N_x/r_x$

$K_y = 2 \cdot \Pi \cdot N_y/r_y$

$K_z = 2 \cdot \Pi \cdot n_z/r_z$

$R \rightarrow \infty \int U_N(r) \cdot U^*_k(r) d^3r \rightarrow \delta N \cdot r$ (Delta Cronnicker).

*****{{For detailed calculations See : Q.M.- L. I. Schiff 1956.}}}

The Unit Sphere (V).

Now we in position to move from the above __SPHERICAL ENSEMBLES__ to give exact and universal definition of the __UNIT SPHERE__ (v) Which I introduced in the above (Master Graph).
From the general static spherical perfect fluid proportional-EOS (equation-of-state)

$p = w\rho$ Except for the trivial cases solution not known yet. is Here we consider the case analytically solvable(Buchanan)(Equation Of State (EOS) parameters w=-1/6

But first we need to keep in mind the followings:

1- We are inside ISOTROPIC universe.

2- The observer is moving with the flux i.e in the X-Direction of the graph above

Thus: The line element for a static spherically symmetric spacetime can be written as

$u\mu = u0 \delta\mu 0 \Longrightarrow u0 = B(r)$

$ds2 = -B(r)dt2 + A(r)dr2 + r2\, d\Omega^{SPHERICAL}2$ Where :

$d\Omega^{SPHERICAL}2$ is the line element for the __UNIT SPHERE__. (v) according to the above graph!

If the spacetime is sourced by an isotropic perfect fluid (SSSPF solutions), We need :

$u^\mu = u^0 \delta^\mu{}_0 \Rightarrow u^0 = B(r)^{-1/2}$.

$A(r)$ and $B(r)$ Are functions not easy to solve such that EFE (Einstein Field Equation) $G_{\mu\nu} = \kappa T_{\mu\nu}$

Is satisfied. for most EoS

3-Cosmic Dust pressure =0.

Thus we cannot use as we did in F.M.5. the EFF

That is to say, the general exact solution of Einstein's Field Equations (EFE) $G_{\mu\nu} = \kappa T_{\mu\nu}$

are not known for perfect fluid (PF) source, if

$$T_{\mu\nu} = (\rho + p)u_\mu u_\nu + p g_{\mu\nu}$$

where ρ(Rau) and p satisfy

$p = w\rho$. $\qquad G_{\mu\nu} = \kappa T_{\mu\nu}$ (1)

with constant w, for the static spherically symmetric (SSS) case. $IG_{\mu\nu}$ is the Einstein tensor, and for its definition we use the conventions : $\qquad \mu = u^0 \delta^\mu{}_0 \rightarrow u^0 = B(r)^{-1/2}$.

κ is the coupling constant, and $T_{\mu\nu}$ the stress-energy-momentum (SEM) tensor.

$T_{\mu\nu}$ describes a so-called perfect fluid, sometimes called the isotropic perfect fluid, which corresponds to a fluid without viscosity and heat conduction, where ρ(Rau) and p are the energy density and pressure, respectively, as measured by an observer moving with the fluid; and u μ is the fluid's four-velocity. Another aspect of the description of a perfect fluid is an assumed relation :

$$f(p, ρ) = 0,$$

called an equation of state (EoS). For analyses of stellar structure, the polytropic EoS, $p \propto ρ^γ$ is often used, while in cosmology, the(isothermal EoS) is relevant. For example, w = 0 describes the matter-dominated (or "pressure-less dust") case, since galaxies are taken to behave like the atoms of a cold gas filling the universe,

_____**Result:**

P=Pressure.

P(rau) = Energy Density.

W=Constant.

$p \propto P$

$p = wρ$

$Ω^{Tilda} \neq Ω^{SPHERICAL}$

Now recall we said that the above (Master Graph)Of Frequmechanic clearly shows that :

n/N=Constant ratio throughout the X-Direction and this was derived from : Rhe volume of each individual sphere divided by Its position on the X-axis is constant:

n.v/N=Constant =n/N at v=1.(Unit Sphere).

Now we translate these facts in the followings:

Energy Density =f($N.h.\omega_g /n,v$)= $N/n \times (h.\omega_g /n,v\_)$

Pressure =f(.h. ω_c) Therefore:

the ratio =$N.h.\omega_g /n,v) \times 1/ (.h. \omega_c) = (hbar/hbar)(\omega_g / \omega_c).N/n,v$

=$N/n (\Phi)$=Constant. Since (v)=unit volume =1

But we already established n/N Or N/n =Constant

After substituting the result will be :

W=Constant = Pressure/Energy density

W = $N/n (\Phi)$= In Frequmechanical terms .

Clearly from the above argument and equations the conclusion will be that the introduction the concept of :

<u>UNIT SpHERE</u> (v)And subsequent constructions In the above (MASTER GRAPH) Of Frequmechanic valid and applicable universally .Most important it contain solution for the Non-Trivial cases of : $p = w\rho$?.

*_____

REFERENCE

{{{EOS parameter w = −1/6 By Ibrahim Semiz Boğazici University, Department of Physics 34342 Bebek, Istanbul, TURKEY }}}.

How To Obtain Unit Radius Vector (R).

Jn all my frequmechanic books I defined **R as radius extending from infinitum. But first we need recollections of some of the equations obtained so far:**

Frequencies $(\Omega) = (\omega\text{cutting}) \pm (J\omega\text{Gluing})$

Angular velocities: $(\Omega.R_{c,g}) = (\omega_c.R_c) \pm (J\omega_g.R_g)$

At: $R_c = R_g$ $R_{c.g} = 1$

$(\Omega.1) = (\omega_c.R) \pm (J.\omega_g.R) = R(\omega_c) \pm (J\omega_g)$ Implying R=1.(UNIT).

At: $R_c \neq R_g$ $R_{c,g} = 0$

$0 = (\omega_c.R_c) \pm (J\omega_g.R_g)$

$(\omega_c.R_c) / (J\omega_g.R_g) \rightarrow R_c/R_g = \pm j.\omega_g/\omega_c$

If we examine the above **Equal Ratio** carefully it imply that:

A- Any change in the frequencies produce extension or contraction of the radii vectors from Infinium by multiples of **R.**

B- $(\omega_c.R_c)/(\omega_g.R_g) = \pm J. = $ Constant.

Now we dente Angular velocities by (OMEGA TILDA)S:

$$(\Omega.R_{c,g}) = (\omega_c.R_c) \pm (J\omega_g.R_g)$$

$(\Omega^{Tilda.}) = (\tilde{\omega}C) - (J.\tilde{\omega}g) \rightarrow$ $J.(\Omega^{Tilda.}) = j.(\tilde{\omega}C) + (\tilde{\omega}g)$

$(\Omega^{Tilda.}) = (\tilde{\omega}C) + (J.\tilde{\omega}g) \rightarrow$ $J.(\Omega^{Tilda.}) = j.(\tilde{\omega}C) - (\tilde{\omega}g)$

Giving Average $(\Omega^{Tilda.}) = \frac{1}{2} J (\tilde{\omega}C)$

Momentum for particle of mass (m) $P = m.(\Omega^{Tilda.}) = m/2 J (\tilde{\omega}C)$.

Probability Density :

1- Although the curve is of infinite length but no matter how far it extend it always return to intersect the X-Axis i.e. Bounded by upper zone therefore the areas enclosed can grow **RAPIDLY** very large yet remains **Finite**.

2- Since the curve itself is continuous describing discrete numbers of areas it justify the smooth flow of probability current density. (No source and No sink) At (X>>4.)

3- Now we are at cross road : Either we can make use of Stoke theorem: For vector function F :

$$_0\Sigma^4 \ _4\Sigma^\infty \ \{ \int \nabla \times n.F \, dA = \int_c F.dl \}$$

However it involve rigorous mathematical treatments therefore it has been delegated to the Frequmchanic SIX.

OR:

Fortunately Q.M. offers short cut to this problem Since we only interested in the expectation value Of the curve intersecting the X- axis from the set values $\{S \ 4^n , 4^n+1 , 4^{(n+1)}\}$

As the areas blows up to infinity not the actual size of the areas itself

By using the all familiar Schrodinger Representations of Momentum (P) For particle mass(m) Keeping in mind that (Omega Tilda)Now represent Angular velocity Not frequency due to Unit VECTOR (R)As defined: i.e OMEGA tilda itselfbecome a vector as follows:

$$P = m \cdot \tilde{\Omega}$$

For particle mass (m) The expansion of any state :

$$p \cdot (R|\Psi>) = p \cdot \{S \ R|\Omega) >< \Omega) |\Psi > \}$$

$$(1/2\pi \cdot h)^{1/2} = \int e^{J.\Omega.R/h} <\Omega|\Psi> d\Omega.$$

According to Classical probability here we have <u>**BIVARIANT DISTRIBUTION**</u> of two discrete functions but fortunately the Quantum <u>**COMPLETENESS PRINCIPLE**</u> simplefy the procedure as follows: First it allow us toLINEAR expand each in to set ;

$$\{S \ 4^n, 4^n+1, 4^{(n+1)}\} \qquad S + |A><A| = 1$$

$$S_{4^n} <\Phi | A> <A| \Psi> \text{ e.g. } A|\Psi> = \Sigma \ A|\delta><\delta|\Psi>$$

(upper,Lower)= Probability current density

$$= d/dt \ (S <\Phi | A> <A| \Psi>) \ \textit{IMPORTANT:}$$

$$A+ = 4\int_0^\infty y \, dx = \int_0^1 F(\Omega) \, dx.$$

$$A- = 0\int_0^\infty y \, dx = \int_0^\infty F(\Omega) \, dx$$

1- _____ __Classical:__

Now referring to the graph above $\quad \delta = F(\Omega)$

And the dentitions for the VECTORS of angular velocities :

$$\tfrac{1}{2} \cdot m\,(\Omega R_{c,g})^2 = \tfrac{1}{2} \cdot m\,[\,\omega R_c + J.\omega R_g\,]^2$$

$R_{c,g} = 1$ at $c=g$ AND $R_{c,g} = 0$ At $c \ne g \rightarrow$ At $R_c = R_g$.

$(\Omega.)^2 / R^2 = [\,\omega_c + J.\omega_g\,]^2 \qquad$ **EQN-1.**

At $C \ne g \rightarrow R_{c,g} = 0$

$[R_g/R_c]2 = [\,\omega_c / J.\omega_g\,]2$ ------------ **EQN-2**

$(\Omega^{Tilda.}) = (\tilde{\omega}c) - J\,(\tilde{\omega}g)$ --------------- **EQN3**

$J.(\Omega^{Tilda.}) = j.(\tilde{\omega}c) - (\tilde{\omega}g)$ ------------------**EQN-4**

_____ __Quantum:__

$$(1/2\pi.h.)^{1/2} = \int e^{\,J.\Omega\ Telda/h}\,<_{d\Omega^T}\,|\Psi>_{d\Omega^T}$$

Substituting from EQN 3 In to the above:

Where all omega-s are now ΩTelda =AngularVelocity Vector.i.e R is embedded.

$$(1/2\pi.h.)^{1/2} = \int e^{\,(J.\tilde{\omega}c)\,-\,(\tilde{\omega}g)./h}\,<_{d\Omega^t}\,|\Psi>d_{\Omega^t}\ \text{------------ EQN4}$$

$$(1/2\pi.h.)^{1/2}\,e^{\,(\tilde{\omega}g)./h} = \int e^{\,J.\omega c/h}\,<\Omega|\Psi>d\Omega\ \text{------ EQN5}$$

OR:

If we substitute for $(\Omega.)^{Tilda} = \underline{R} [\omega c + J.\omega g] \rightarrow$

$(1/2\pi.h.)^{1/2} = \int e^{(\tilde{\omega}c) - (J.\tilde{\omega}g)./h} <\Omega|\Psi>d\Omega$ EQN 6

$(1/2\pi.h.)^{1/2} e^{(J.\tilde{\omega}g/h} = \int e^{\omega c../h} <\Omega|\Psi>d\Omega.$ EQN7

Multiplying EQN 5 X EQN 7 $\rightarrow$

$(1/2\pi.h.) \quad e^{(J. + J)\tilde{\omega}g/h} = \int e^{\omega c(1+J)./h} <\Omega|\Psi>d\Omega.$

$(1/2\pi.h.) = \int e^{(\omega c + \tilde{\omega}g)(1+J)./h} <\Omega|\Psi>d\Omega.$

Now we have the expectation value for R JUMPED by one unit (1)Digit as one unit Radiue vector from the above result. ALSO: If we now divide EQN 7 by EQN 6 we get:

$e^{(1- J)\omega g./h} = \int e^{(J-1)\omega c/h} <\Omega|\Psi> d\Omega.$

$e^{\{(1- J)\omega g - (J-1)\omega c\}/h./h} = <\Omega|\Psi> d\Omega.$

$e^{\{(1- J) (\omega c + \omega g\}/h} = <\Omega|\Psi> d\Omega.$

Probability Current Density.

"Here we prive scientifically the usefulness of the Cosmic Flux concept.
"

$d/t \ (Prob <\Omega|\Psi> d\Omega. = (1-J)/h \ e^{\{(1-J)(\omega c + \omega g\}/h}$

But first we need to introduce 3D Probability P(r,t):

$d/dt \ Prob \ (r.t) = \int space \ [\ d/dt \ \Psi^* + d/dt\Psi]d3r.$

$=ih/2m = \int s \ [\ \Psi^{**} . \nabla^2 \Psi - (\Psi^* \nabla^2)\Psi \ ] \ d3r.$

$=ih/2m = \int s \ \nabla [\nabla \Psi^{**} - (\nabla \Psi^*)\Psi \ ] \ d3r.$

$=ih/2m = \int s \ \nabla [\nabla \Psi^{**} - (\nabla \Psi^*)\Psi \ ] \ d3r.$

Now by using Green Theorem :

$=ih/2m = \int A \ [\nabla \Psi^{**} - (\nabla \Psi^*)\Psi \ ]n \ dA.$

n=Component of the vector normal to the direction of wave packet.

Introducing Component of the vector in the directin pf the normal to the surface dA. vector:

$$\mathbf{B}(r,t) = [\nabla \Psi^{..} - (\nabla \Psi^*)\Psi]$$

$$d/dt \int_s \text{Prob}(r.t)d3r = -\int_s \nabla \cdot \mathbf{B}\, d3r = -\int_A \mathbf{B}n \cdot dA$$

And the probability current density is :

$$dP(r,t)/dt + \nabla \cdot \mathbf{B}(r,t) = 0$$

This lead to the following conclusion:i

$-h/2m\, \nabla$ Can be identified with MOMENTUM. Thus the velocity operator $= (h/i.m).\nabla$ Giving in terms of Frequmechanics→

$B(r,t) =$ The real Part of $[j\, \tilde{\omega}g \cdot (h/i.m).\nabla \tilde{\omega}c]$

Where $\tilde{\omega}c$ & $\tilde{\omega}g$ are the omega tilda representation of

UNIT ANGUKAR VRLOCITY i.e Divided by (n)?

Clearly OMEGA tilda is function of time Now by differentiating through the above with respect to TIME expressin we get the PROBABILITY CURRENT DENSITY

$$d/t \langle \Omega | \Psi \rangle \, d\Omega. = (1-J)/h \; e^{\{(1-J)(\omega c + \omega g\}/h}$$

Result is that the the process of cutting and gluing the C.F is flowing similar to that of no source and no sink current justified by the fact there are no sources or sinks since the curve assumed to run smoothly to infinity.

Clearly by Multiplying EQN- 1 By h^2 (h=Planck Constant) Then cancelling it from both sides shows that this equation one for energy holds in both quantum and classical case
Same can be said of EQN-2 Notice that all the Ares Subtended by the curve in the Positive Field between h and (h+1) Are equal =

$$\int_0^1 F(\Phi) \, dx.$$

And the total sum of the areas in the positive Quantum Field obtained by multiplying with the expression for the discrete numbers (N):

$$A_+ = 4\int_0^\infty y\,dx = \int_0^1 F(\Omega)\,dx.$$

$$A_- = \int_0^\infty y\,dx = \int_0^\infty F(\Omega)\,dx.$$

And this what is behind our ability to normalize functions ?

2-
The Areas below the n.h -Axis are not equals and <u>Rapidly Blows Up Asymptoticly To Infinity. And This Is How We Can Terminate Infinities At (Renormalization).</u>

3- <u>There fore it can be said For Most Earthly Calculations that infinity at $(4)^n$ when $n > 10^B$ (ORDER OF MAGNITUDE) as defined in this Frequmechanic Three..</u>

<u>4-</u> That all the areas in the Positive Quantum Field remains constant CONFIRMS my SEARCH for UNIVERSAL CONSTANT(H) Described and Derived in all my other Books?

Hence if we multiply EQN-! By h^2
OR by my universal constant $(H)^2$ <u>The result will be the same</u>

5-

Moreover EQN-2 : $[R_g/R_c]^2 = [\omega_c/J.\omega_g]^2$

If we multiply one side by (h)And the other side by (H)The results are the same by applying the analogue of the equations for the areas subtended by the curve in the POSITIVE Zone of the diagram above.

From EQN-1

$$E = F(\Omega) . [\ _0\int^a dx.4^n - 4\int^a m.\,dx] = h^2[\ \omega_c + J.\omega_g\ ]^2$$

From EQN-2 At $c \neq g$ → $[h.R_g/H.R_c]^2 = [h.\omega_c\ H./J.\omega_g]^2$

By substitution and cancelling it can be seen that ;

h→HAnd H→h For all expressions Classical or Quantum!

Its likely that H is multiples of h the question is what are these multiplication factors ?

Are they constants themselves or operators of variable OR combination of both.With present knowledge available to us this can be only deduced and verified at the same time ONLY empeically.

6-Mathematically since the value of the total sum of areas in the lower zone rapidly increases but still remain **FINITE** and bounded in evaluation by the areas in the upper zone .

But there is no equality ONLY (ORDER OF (β)As defined above.

7-
Due to the huge values of the summand any digital change in (n) will have negligible change on the SUMMAND therefore its this fact allow us to exchange the summation operation with that of integration.

8-
To put my IDEA in nut shell :
What we are doing is to introduce a curve with the following properties
1-infinit length.

2=Smooth Differentiable.

3- Criss Crosses the X- Axis at Discreet intervals.

4- escribing equal physically measurable areas above the X-Axis..

5 -To shadow OR (keep track)Of Areas blowing to infinity Below the x-Axis while the Y-Axis apply the Quantum postulate of the <u>POTENTIAL WELL</u>;

Brief Extracts From My Other Nine Books ?

$J = \sqrt{-n}$ $n = 0 - \infty$ Alpha $\alpha = 10^{-\beta}$ (Minus β) $\beta = 0 - \infty$.

1- $\Omega = \omega\text{cutting} \pm J\omega\text{Gluing}$

1A - $.\Omega^2 = (\omega C + J.\omega g.)^2$

$.\Omega^2 = (\omega c)^2 - n(\omega g.)^2 + 2\sqrt{-n}(\omega C . \omega g.)$

Therefore at resonance $\omega c = \omega g = \omega$

$.\Omega^2 / \omega^2 = (1 - n + 2\sqrt{-n})$

Similarly :

1B- $.\Omega^2 = (\omega c)^2 - (\sqrt{-n}.\omega g.)^2 - 2\sqrt{-n}(\omega C . \omega g.)$

$\Omega^2 = (\omega c)^2 + n(\omega g.)^2 + 2\sqrt{-n}(\omega C . \omega g.)$

At Resonance :

$.\Omega^2 / \omega^2 = (1 + n - 2\sqrt{-n})$

Solving 1A = 1B·

$(1 - n) + 2\sqrt{-n} = (1 + n) - 2\sqrt{-n}$

$2n = 4\sqrt{-n}$ OR $4n^2 = -16n$ · $n = -4$

AND:

$2\Omega 2/C\Omega 2 = (1 - n + 2\sqrt{-n}) + (1 + n - 2\sqrt{-n}) = 2(1+\sqrt{-n})$

After substituting for n=-4 above $\cdot \Omega 2/C\Omega 2 = 3$

1- Thus $C\Omega = \sqrt{3} \cdot \Omega$ This is more important result than what it seems since it justify taking (CΩ)As the quantum average by taking in to account the three states of measurements for

(Ω, ωcutting, ωGluing)

In this theory object are identified simply by their phase or lag angle i.e. absolute number where :

œ = | Φ |

RESULTS:

1- Gravitation:

Applying the previous example calculating alpha for black hole (ɑ)Blackhole and (ɑ)Apple To obtain the gravitation betwee the two we need to insert these two values of ALPHA in to Tensorial form of my Equation: Substituting :

ɥ=(œ)Blackhole and V=(œ)Apple

In Tensorial form

$\Omega^{\mu} \cdot v = (\tilde{\omega}C)^{\mu} + j \cdot (\tilde{\omega}g)_v$

The Equivalence Principle And The Relative Phase:

"The choice of Apple and the Blackhole in these examples was not ad hoc only to emphasize that although mass of apple can be negligible compared to that of the blackhole yet my method can take both in to the calculations. Because although the value of the phase angle is between (0 and 1) Its range $\beta=(0$ to ∞)"

1-
To find the gravitational energy between the apple and the black hole First find third object EQUIVALENT In Phase Identity to the combined phase identities of the Apple And the black hole thus:

Relative Phase = (œ)Blackhole /(œ)Apple.
Which is in the example above = œ(61) = 10^{-3} / 10^{-64}

Gravitational ACTION = $(10^{+61}.)$ $(H\mu - H\nu)$ erg.sec.
REcall : Force = $\nabla \nu$(Potential Energy)

2-
To find the Combined Mass (Energy) Of the apple and the black hole First find third object EQUIVALENT In Phase Identity to the combined phase identities of the Apple And the black hole thus:

Relative Phase = (œ)Blackhole X (œ)Apple.
Which is in the example above = œ(67) = 10^{-3} X 10^{-64}

Combined Mass ACTION = $(10^{+67}.)$ $(H\mu + H\nu)$ erg. sec.

My Universal Constant (Hn) Also Fifth and Sixth Isamic Deltas (δn & δm)

$$H_n = (½)[h_n + \delta_n \cdot (h)\delta_m] \cdot e(\omega_c + \omega_g) \cdot S$$

Now for ($\approx$) · (=) Let: $S = (n.h)\delta$

At n=0 ? δn =1 and at n=1 ? δn =0 At >1 δn = +1.

AND: δm = -n S·(n.h)-n.

Giving: $$H_n = (½)[h_n + \delta_n \cdot (h)\delta_m] \cdot e(\omega_c + \omega_g) \cdot S$$

$$H_n = (½)[h_n + \delta_n \cdot (n.h)-n] \cdot e(\omega_c + \omega_g) \cdot S$$

h = Planck Constant (erg.Sec) !

H = My Proposed Universal Constant (erg.Sec).

Ω = Hybrid Frequency made of two frequencies ω_c & ω_g!

n = 1,2,3,..etc

φ = Relativistic ally calculated Phase angle By:

See Previous section.

Intrinsic Probability.

"This section will turn Q.M. Upside Down on its head where the \UNCERTAINITY PRincile is TACTICALLY VIOLATED but STrATUGUCALLT maintained."

" Between Zero and one there are infinite numbers of WELLS and SLOTS each determining the probability of finding from infinite number .of mcroscopic or macroscopic object specific physical object."

"Its meaningless talking about the pribabilty of finding neutron in neutron star we all know the answer is one!Its more meaningful asking about the probability of finding Neutron Star in the Universe/"

based on the exclusion principal of Q.m.That no two particles can occupy the same state and that the range:

(β) ==(Zerp to Infinity of the T.F.

(œ) =(Zerp to 1y) There are infonit number of slots between Zero and one.

This section attempt to prove that the T.F (transfer Function (œ)= Zero to One has Infinite number of slots can be taken as the intrinsic probability of each particle which is (Zero To One)thus particles relativistic or otherwise can be identified by these T.F ≡ Prob:Thus

Paricles such as Liptons situated nearer to one while macroscopic objects are towards Zero.

However caution is required with relativistic particles photon since it has multiple numbers as its absorbed and generated by encounters.

The assumption that its more than coincident the range of the T.F.( )Is the same as the classical probabilies can only be justified experimentally and

when this happen it can be taken as one of the evidence for the existence of the cosmic flux.

$(œ)n =$ (Zero to One)   Intrinsic Probability for particle (n).

$\beta = $(o-∞) Range of slots

As before $(\tilde{a}) = 10^{-(Minus)Betta}$ Delta $= \Phi / 2\pi$

delta x= delta $\Phi . \lambda / 2\pi$ Delta x.deltat = ?

The Emiton/

"According to Eeherfest theorem that in slowly varying potential field moving wave packet can be represented by classical particle which I named the EMITON

In the classical limits of quantum mechanics, its where Frequmechnic can be best understood.

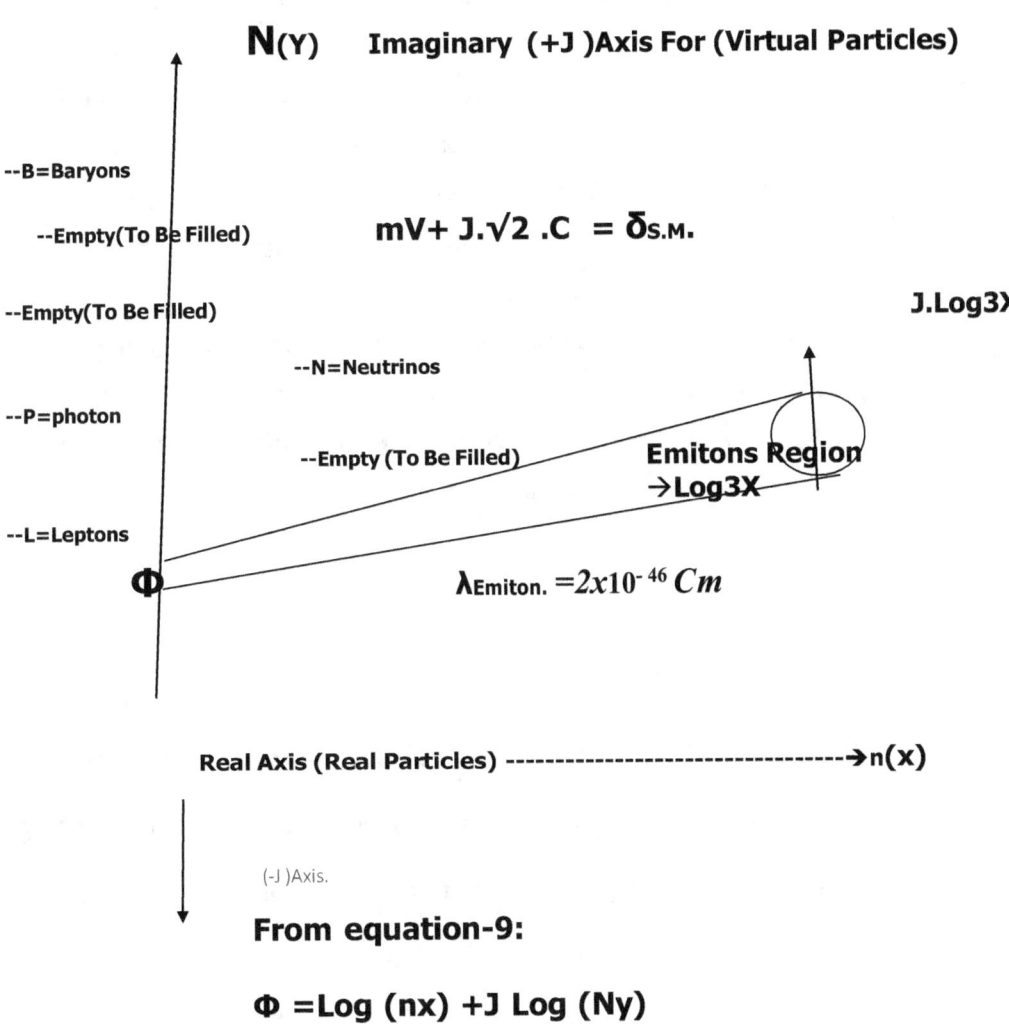

From equation-9:

Φ = Log (nx) +J Log (Ny)

ON NEGOTIATIONS:

" Is it possible ? To obtain mathematical expression for the process of negotiations by operating in reverse! By backtracking! By designating zero to <u>Failed Negotiations</u> thendeveloping the expression backward" We all trying to be clever? !All the negotiators are trying to beclever by Maximising their positions in pretending position -a-is really –a"- b,is really -b"   And C" instead of C etc. But the truly clever are those who try to be <u>Less Clever</u> !

<u>Henry Kissinger</u> Must have discovered this formula <u>Empirically</u> (By Practice)Because he was in the habit of making <u>Pre Emotive Offers</u> sending his opposites reeling surprised and destabilised such that it made it easy for him to go for the kill."

We all trying to be clever? ! Too Clever !All the negotiators are trying to be clever by maximising their positions in pretending position -a- is really –a"- b,is really -b"   And C" instead of C etc?

Let us say we have two negotiating teams:

The Romans "R" (appropriately denoted by Roman letters).

with their positions a, b, c, etc **Pretended** to be a", b", c".

And negotiating some other team:

The Greeks "G" over some dispute say **border line a curve?**

Inside an Island e.g. Cyprus (their positions appropriately denoted by the Greek letters).

α, β, γ ---------pretend**ed** to be -------------- → α", β", γ".

The Romans will try to maximise their positions by:

$a''-a = Д_i$ $b''-b = Δ_j$ $C''-C = Δ_K$ ETC.

And let the maximising for the Greeks be:

$α''-α = δ_i$ $β''-β = δ_j$ $γ''-γ = δ_k$ ETC.

i.e. One Border Line Is Represented By α", β", γ". While The Other Described By a", b", c".

_____ **HINTS.**

"Entrust The Maximizing To The Maximizers?"

NOW FOR UNSUCCESSFUL NEGOTIATION:

$Д+δ=o$ $\qquad Д = ± δ$ (stalemate) etc.

SUCCESSFUL NEGOTIATION

$Д+δ=n$ giving $\dfrac{Д+δ}{n}=1$

I.e. $Д/n + δ/n = 1$. The trick is to find "n" which is Unfortunately must be a whole number?

Now let $m ≠ n$. i.e.

$Д/m + δ/n = К P$ where $К = I, j, k$ etc P for prime? Hence:
$Д/mК P + δ/n КP = 1$

This imply that $\qquad mК = nК \text{-------→} \quad n = m$

*(Please note that in the Latter case: - a,b,c, and a",b",c"etc are

the **Positions?** Both "real and pretended" by the negotiators

over a range of issues and no longer members of any community as they were in the first case?)
***AGAIN** PLEASE KEEP IN MIND THAT THIS IS A **CONDENSED**

PRESENTATION and introductions?

Now although (R) and (G) are opposite each other (physically) it's the (R") and (G") that are negotiating!
The questions now; will the following expression hold true?

(Morphology.) $\quad \Sigma \, Д_{I,J,K.} \; -- \; \Sigma \, \delta_{I,J,K.}$

$$R" - G" - (R-G) \quad \approx \; \text{---} \; ? $$
$$n$$

***Hint:**Could "n" be the number of **Sessions** per negotiation?

And that the critical session falls at : $x = [\,^{r}\!\sqrt{n}\,]$?
where "r" is the number of **Negotiators** not negotiations?

r > 1. [EVEN or ODD]

R=The number of negotiating teams. [EVEN]

Where   Rx r = N.

Also n-must be whole number perfect square in the case 0f r=2 or

x=whole number in other cases.(see above).

More hints ?

If By definition:

"R"   must be a whole number [EVEN]?

And

" r" also must be whole number [Even or Odd]?

Therefore :

N= will be <u>Both</u>: WHOLE NUMBER &EVEN.

$n^{1/r}$ = Also must be a whole number .

[All of these will be discussed at length in my next book thatwill be strictly. Mathematical)

The Constant Of Destructivity.

"Let us now take my philosophy on the <u>Constant Of Destructivity</u> Further in to real physcs:Based on the <u>Exclusion Principle</u> of quantum mechanics which states that :No two quantum paticles can occupy the same state.

The Comparability Principle And Index :

"In our calculation inside the <u>Macroscopic World</u> we stay with the <u>Exclusion Principle</u> of quantum mechanics except now we add the following (Comparability Principle) Stating that ((Inside any ensemble of zones $Z_{A,B,C,D,etc.}$ Although no two zones can occupy the same state yet any two can be comparable if their states are comparable))(Not to be confused with the <u>Complementary</u> principle in quantum physics)."

Imtroducing The Comparability INDEX (Σ):

"How to determine the <u>Nearest Two</u> Zones ?"

Theis exercise is clearer in particle physics than in the macroscopic world since each <u>Ensambles Of Zones Or Fields</u> of quantum particles Z(A,B,C,D,etc)? Has well defined specifications e.g. Mass?Spin?Charge?etc !Hence they can be classofied in specieis !Not so in the Macroscopic Classical World.

Thus let us take this further in to the macroscopic world of ensambles of where the situation is less defined and requires <u>Careful Analysis</u>.

For simplifications we choose <u>Geopolitical Zones</u> such as Britain And Germany since theyse are the most faniliar examples to understand however the <u>Procedure Of Determining The Nearest Two Out Of Any Ensemble</u> is not so easy requiring numerous steps to exclude other zones from the <u>Nearest Two</u> for which the following criteria in steps may be helpful in determining the <u>Nearest Two</u> in other spheres e.g. Biological or Chemical zones etc.:

Now before proceeding further we need to determine (In Practical Terms)The Comparability INDEX (Σ)AS follows :

Our choice of Britain and Germany was not arbitrary:
Apart from the **Proximity** of these two zones in many fields these two also are most familiar for us :
we could have chosen any two (K&R)Zones if they display the nearest qualities:
Here is **Guid Sample** of how to proceed ?A list of points of what should be considered in order of their importance?

1- *Both have or had at some point similar **Population Density**.*
2- *Alhough their two climates are not identical yet they are the **(Least Different)** in Europe.*

3- *Both largely Anglosaxons*
4- *Both had abandoned the christan church (Catholicism) For the wilderness of protestanism.*
Both Britain And Germany Committed Massive Gemocides Of Jews !

(Search My E-Books---------- The First Genocide)?Also:
(Search My E-Books---------- The Enemies Of Mankind)?

5- *Both never had **Successful** Revolution.*

6- The __Relationship of Production__ Inside both had reached its utmost vulgarity thus both suffer from __Unique Eerie Totoal Absence Of Social Life i.e__ their people had been dehuminized in to __Robotics__ (Compare France)

6- Both suffer from __Superiority/Inferiority__ Complex therefore they Nestle Deep Seated __Incurable Disrespect__ for others ! How sad and hard it must be for any scientificly minded person devoted to none violence to admit tumbling on hard to ignore __Scientific Finding__ that such __Chronic Disease__ Had become __Incurable__ It caN only be corrected by force of arms?I am afraid this __Chronic__ Superiority/Inferiority Complex is not just the catalyst but one of the main ingredient for the __Constant Of Destructivity__.

7-

"The __Methods__ can be as __Sneaky__ as that of the __Average__ English But the __Results__ always will be as __Visible__ as those of the NAZIS."   **The Red Monk.**

In spite of all appearances to the contrary both Britain and Germany are sitting upon <u>Immovable Sub-Terrain Cult Subconsciously Worshipping Racism</u> As if <u>Race</u> was the final and only truth!

Exteremly serious point because it's the <u>Primary Motivating</u> Force Behind both ($D_{colonialist}$ And $D_{fascists}$) Destructivity.

All such scientist can do is to warn everyone that in spite of all the sacrifices and traumas of the second world war the <u>Demons Of Fascism</u> inside European souls such as the <u>English German Or Ukrainians</u> Have not yet been fully <u>Exorcised</u>. And its called the <u>Constant Of Destructivity</u> .

One ukrainian communist told me how it was when he was just school kid a German army column was passing outside his town in en route to Russia .

Most of our youth rushed to welcome them with flowers or sweets as our <u>Saviours From Stalinism</u> !

I was lucky my bicycle broke down while crossing narrow farming canal because when I arrived at the

scene the German army had already passed through but leaving behind corpses of young people some were my own classmates each hanging from each telegraph pole some had under their dangling feet the very flowers they did bring to welcome(Our saviours from stalinism) To spread terror in the population the _Germans_ attached pre printed posters to the deads stating in German And Russian (Do not remove !this will be the fate of anyone who disobey the _Third Reich_).
After witnessing all of this I joined the _Communist Youth Resistance_ But now decades later I still get nightmares of how can people do this to each other?Said he.

This Ukranian Communist reminded me of my own nightmares when I count the number of innocent kids overseas students who arrived to England on _Official_ scholarships to study here only to be (Sneakly) and slowly Poisoned to death or to permanent disability by _Fascists English Circles (With Full Blessing And collusion Of The State)_!

Often for no better reason than (Their faces did not fit in) Or the said overseas student had been rude to advances made by an <u>English State Homosexual</u> etc ! ? If you do not believe all of this I leave you with one word (AZOV-2022). That is Azov (the Anglosaxons).

"The <u>Methods</u> can be as <u>Sneaky</u> as that of the <u>Average</u> English But the <u>Results</u> always will become as <u>Visible</u> as those of the NAZIS."   *The Red Monk*

The Science :

So far these two Britain&Germany scored seven points thus $\sigma = 7$. Other comparabe zones may score less or more points By now it should be clear how we can evaluate the <u>Comparibality</u> factor here for Britain&Germany:

$\Sigma = 7$ Thus:

$$Z_{Colonialists} - Z_{Fascists} = \triangle / \Sigma$$

Where Z=Zone   t=Time

D=Destructive <u>Force</u> Measured only by the number of people killed since its the only available scale more reliable than that for example destroyed properties etc.

D_c = Destructivity of Colonialists

D_F = Destructivity of   Fascists.

Z_c = Colonialists Zone Z_F = Fascists Zone.

Δ = The Difference Facror.

Σ = The Diluting operator of the differences .

Clearly as $\sigma \to \infty$ Then $Z_c \equiv Z_F$

And if $\Sigma \to 0$ then $\Delta/\Sigma \to \infty$ i.e (Disjoint)

<u>Applying The Laplace Transformation</u> :

$F(t)_{(Colonialist)} = {_0\int^\infty} e^{-D_c \cdot t} F(t) \, dt$

$F(t)_{(Fascist)} = {_0\int^\infty} e^{-D_f \cdot t} F(t) \, dt$

$Z_c = {_0\int^\infty} e^{-D_c \cdot t} F(t) \, dt$

$Z_F = {_0\int^\infty} e^{-D_f \cdot t} F(t) \, dt$

And the difference between these two is minimal (Δ)

$$Z_c - Z_F = \Delta/\Sigma$$

"Done in reverse (By elimination):Since I am doing <u>Quantatative</u> analysis it permits comparing relatively comparable entities by <u>Excluding</u> all other zones than the two possibly nearest e.g in size !Age and stage of industrialization!Religion (Both largey notcatholic) Race (Both not Latin)etc!etc!Any thing (<u>Qualitatively</u> not near enough)Will surface inthe final analysis and can be trimmed or discounted without effecting drawn conclusions ." **The Red Monk.**

The equation for the constant of destructivity Appearing in this book in many shapes forms and applications, Starting from the vague and primitiverelationship between four variables:($D_1 \times T_2 = D_2 \times T_1$: To the better defined :

$$.t \int T \ D_{fascist} \cdot dt = t \int T \ D_{Colonialists} \ dt$$

$$1^t \quad (\quad 1^t \quad (.$$

Now we are in position **to <u>Express It Graphically</u> By takingtwo <u>Relatively</u> Comparable entities? Comparable by <u>Size ! Industry ?Per Capita !Race</u> !etc.)Such as <u>Nazi Germany</u> and <u>Colonialist Britain</u> as follows: <u>Destructivity</u> =There are all kinds of destructivity' Material! Psychological! Environmental etc But The most <u>Reliable Yard Stick</u> We can have for** *measuring destructivity is a number ?The <u>Number of People Killed</u> !These are <u>Concrete Figures</u> And the best there is <u>Recorded And Documented</u> .*

<u>The Period</u> =*The tine that particular destructivity had taken e.g. !*

<u>Six Years</u> = *The period for <u>Nazi Germany</u> Had taken From the day it occupied Poland to the day <u>Berlin</u> Was Liberated by the <u>Red Army.</u>*

<u>Two Hundred Years</u> = *The period for <u>Colonialist Britain</u> Spanning(Geographically) From the day <u>Captain Cook</u> and his (Explorers)Landed in <u>New Zealnd</u> Massacring Unsuspecting natives:*

(See Page—Nature Knows Only Vengeance) ?
Passing through the <u>Countless</u> Massacres the British Committed in <u>India:</u>
(See Page---The Lal Bagh Temple Massacre) ?
To the <u>Thirty Five Millions</u> Perished during the partition of India ? To the numerous massacres they Committed in the <u>Middle East</u> Even at a wedding:
(See Page---Dirty Colonialists Part-32)?

When the English opened fire on <u>Hundred Iraqis</u> <u>Including Dozens Of Jewish Musicians</u> During Wedding ceremony in Baghdad (Documented)!
To the nearly <u>Ninety Millions</u> killed in Africa !
To the invasion of Iraq in (2003) ?
All are <u>Real Figures Recorded By Real Documentation</u>.
Now let us draw the graph with ;

<u>Vertical Axis</u> :

Representing the <u>Rate Of Killing In Million Per Year</u> On this axis an <u>Arbitrary Scale</u> From <u>One To Hundred</u>. Now let us be <u>Very Mean</u> with <u>Nazi Germany</u> Giving it the <u>Highest Rate Of Killing</u>
<u>Rate Of Killing</u> =One Hundred millions Per year .
And let us be <u>Very Lenient</u> with <u>Colonialist Britain</u> Awarding it only a Rate of killing = Threemillions/year.

<u>Horizontal Axis</u> :

Representing the periods in years as described earlier
Thus the area (abcd) For <u>Colonialist Britain</u>
= 3X   200   =600   (Rate   .Year)

=*the area (ABCD) for* <u>Nazi Germany</u> = *100X 6 = 600(Rate .Year).*
<u>These Equal Areas Is The Constant Of Destructivity</u>. *Area(abcd)*
=*Area(ABCD).*

<u>Applications.</u>

"*There is only space here to quote the following application out of many more?*"

Once the <u>Constant Of Destructivity</u> *Evaluated (In this case its the area* =*The figure of 600) for any* <u>Reasonably</u> *Comparable two ? then*

we can obtain <u>Numerous</u> *fundamental information as Follows:*

1-From the inverses of (**Dc & Df** *)* **We can see how the Constructive Contribution** *of the Germans still* **substantial while that of the English approaching zero.**

2- We can estimate the number of <u>Crominal Genes</u> *in each Zone.*

3- **We can predict with** <u>Reasonable</u> **Accuracy the** <u>Future</u> **Behaviour of one entity from the** <u>Past</u> **behaviour of Another !**   **The Ensemble Z**

4- _____ <u>(A,B,C,etc):</u>

By forming a <u>Chain</u> *of* <u>Reasonably Similar</u> *Observable entities*
  we can obtain <u>Limitless</u> *Forecasts from* <u>One To</u>

The Next Once the area is *Fixed (As above =600)* A *Common Constant Of Destructivity* Can be evaluated for the whole *Ensemble*!

5- **The graph is telling us in <u>Concrete Terms</u> that <u>British Destructivity</u> is at least <u>Five Times</u> that of Germany Calculated as follows :**

To take the <u>Average</u> we need to <u>Rescale</u> The Vertical axis By taking the <u>Square Root</u> of each rate of killings Thus <u>Nazi Germany</u> will be at $\sqrt{100} = 10$.

While for Britain $= \sqrt{3}$

Multiplying these two by the periods gives the following : 10 X 6=60 Million Which is exactly the number of lives lost due to Nazi occupation (of Poland ? the USSR? Eastern Europe including Germany itself.etcE

For <u>Colonialist Britain</u> $= \sqrt{3}$ X200 $\approx$ 240 Millions Here we need to take the following in to account: All the well documented evidences and recoded numbers of people killed by <u>Colonialis Britain</u> Add up to about (150 Millions)i.e. Less of what we obtained from the graph and this can Only be explained by the fact

that in third world colonies due to **Socia Conditions** if one person is killed the whole family is killed especially in **Times Of Peace** unlike Nazi killing taking place (In Times Of War) Or **At Least** Two persons e.g. The person plus their spouse ! Therefore this figure of (150 millions) In reality **Double** the figure of (150) ≈ 240 Million killed.

Doubled To approximately (250millions)! **Remember** Since we are here discussing destructivity and not just absolute Numbers. Hence:

The Destructivity (**Not Rate Of Killings**) = 240/60 = 4. **I.e. Contrary To All Appearances English Destructivity Is At Least Four Times That Of The Germans !**

6-
Number Of Criminal Genes Mutated Inside The German DNA ÷ (Divided By) Number Of Criminal Genes Mutated Inside the English DNA = **K.** 6 /200 Years That Is To Say : Number Of Criminal Genes **Mutated** Inside The Germans ≤ 1/30.

Which Is <u>Minute Fraction</u> By Comparison To The Number Of Criminal Genes inside the English.

It's important to observe this concept Of :

<u>Absolute Recirculatiry</u> Nature of crime ?That is to say no matter how any crime against fellow human is packaged (By Nationalism? Socialism?? State ??? Security ???etc!etc)

<u>**Contrary To All That We Wish It Just Does Not Go Away**</u> weather its committed consciously or unintentionally.

But as <u>Negative Energy Changing From One Form To Another</u>.

it simply changes from one form to another. Period.

(See Pages-----Criminal Genes In Action)Parts 1-72) ?

* *(The injection of my name (Isamic)Driven not by any <u>Egoism</u> But by the*

fear of losing my intellectual rights <u>Again</u> and <u>Agan</u> .)

The Isamic Law Of Destructivity And Constructively.

"Its customary to present definitions before not afterany hypothesis !Here I am doing the opposite for obvious reasons of simplification."

We already assumed that destructivity most reliable <u>Scale</u> is the number people killed (D)Therfore its reasonoble to <u>Scale</u> constructivity (C)as its inverse i.e C= (σ/D)Keeping in mind we are only talking about arbitrary scale and Δ is diminishing differences then*It follows:*

$D_1 - D_2 = \Delta D$

$$C_1 - C_2 = \Delta = \frac{(\sigma/D)_1 - (\sigma/D)_2}{}$$

$$D_1 - D_2 = \Delta = \frac{(\sigma/D)_1 - (\sigma/D)_2}{\sigma \quad \sigma}$$

$$= -\sigma[(D)_1 - (D)_2]/D_1.D_2$$

If Costuctivity −Destructivity $= \left(\sigma/D - D\overline{\sigma}_\Delta\right)$ Then we can also say that:

$D^2 + D.\Delta D = \sigma$

Similiarly:

$C^2 + C.\Delta C = \sigma$

$$D^2 + D.\Delta_D = C^2 + C.\Delta_C$$

For any system of conservation:

$$\Delta_D = \Delta_C = \Delta$$

Giving:

$$D^2 - C^2 = \Delta(C-D)$$

$(D-C).(D+C)/-(D-C) = \Delta$

D+C = -Δ ---------------- EQN-A.

Interpretation of -EQN-A

True only if both C&D Are Negative? 2-At diminishing (-Δ.)^n It depends if : n=Odd Number OR n=Even Number ??
And This Is My Law Relating The Two By Constant(σ) The <u>Crux</u> Of My Idea That Any Differences In Destructivity If the Term Is Positive It Can Be Taken In Practice As <u>Constructive</u>.(Think of Better OR Worse?)<u>i.e its better to have (D2)than (D1)(the lesser evil)etc.</u>

Equal or similar entities with <u>Different</u> Destructivity Or constructively sweep equal areas called the constant ofdestructivity. As shown in graph A1.

"The most skirling potential to this theory of mine it's when we add <u>Another Dimension</u> e.g. the population density of the <u>Killer Race</u> AND/OR that of the <u>Killed Nation</u> !!

Such that we obtain equal areas from the <u>Multidimensional</u> Picture by the projections of various orientations!!Meanwhile we try to keep it simple as follows:

Looking at the <u>Positive Side</u> By Applying the <u>Inverses</u> of these <u>Ratios</u> We discover the <u>Constructive</u> Side Which <u>Tally (Fully Agree)</u>with all our <u>Practical Honest Observations</u> Be it in German sciences! Technology! Arts (Especially Music Or sculpture) Philosophy? Etc ?? The <u>Originality</u> of <u>German</u> Constructivety is <u>At Least Four Times</u> that of Britain's <u>Fake</u> Constructions !!There are several versions to this method for example if we equate both periods say to <u>200 years</u> and plug in the figures of people killed by each side we simply find the area (ABCD)Is now stretching flat on the X-axis keeping same value but in <u>Different Orientations</u> etc?!?

(See Parts --- Nation Of Warriors) ?(See Page --- Herschel)?

How Herschel The German Astronomer was Hi Jacked By The English??Also:(See Page-Isaac Newton)?

How Isaac Newton Was of Of Dutch Not Engl;ish Origin??Etc

_____Exercise:

1-It's not difficult to see though <u>Mathematically</u> Both areas are equal! Yet <u>Geometrically</u> one area (abcd)Is <u>Time-Like</u> While the other is <u>(Rate)-Like</u>.

1- Why have we taken the <u>Square Root</u> of the values on the Y-Axis but not those of the X-Axis ?Has this anything do with <u>Normalizing Probabilities</u> ??

HINT: $(-\Delta.)^n$

Rate of killing millions Per Year.

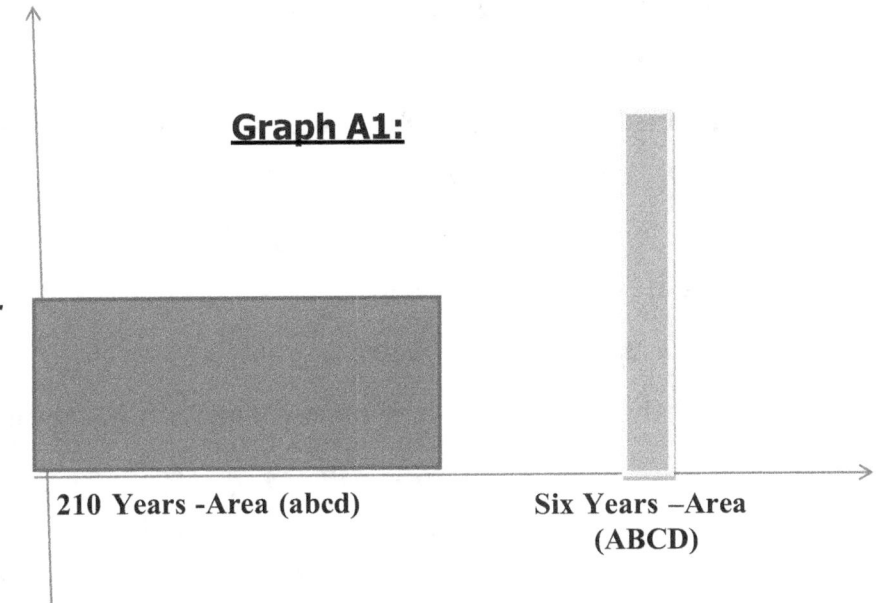

Graph A1:

210 Years -Area (abcd)   Six Years –Area (ABCD)

The Y-Axis:
10 =Rate of killing Per Year by nazi Germany .
3 = Rate of killing Per Year by colonialist Britain.

The X-Axis:
210 Years =The period for colonialist Britain
6 Years =period for Nazi Germany.
"The two Yellow and blue areas (ABCD)And (abcd)Are equal due to the constant of destructivity" **The Red Monk.**

Advanced Version Of My Equation On The Constant Of Destructivity:
As defined earlier: Applying The Laplace Transformation:

$F(t)_{(Colonialist)} = \int_0^\infty e^{-D_c \cdot t} F(t) \, dt$

$F(t)_{(Fascist)} = \int_0^\infty e^{-D_f \cdot t} F(t) \, dt$

$Z_C = \int_0^\infty e^{-D_c \cdot t} F(t) \, dt$

$Z_F = \int_0^\infty e^{-D_f \cdot t} F(t) \, dt$

And the difference between these two is minimal (Δ)

$$Z_C - Z_F = \Delta/\sigma$$

"Clearly $Z_C = Z_F$ when $\Delta = 0$ OR $\sigma \to \infty$ Then by back tracking methods I established that my <u>Comparability Operator</u> (Δ/σ) Can replace the limits of the Laplace integrations and subsequently replacing the resulting constant of their indefinite integration."

Now let us plug in some figures we are relatvely confident of and see what happens : <u>We are sure that:</u>

t_F = six years : Period (t) for German destructivity (the second world war period)

T_{Col} = Two Hundred Years : <u>Minimum</u> period for British Colonialism.

We also confident of the figure for German destructivity (D_F) = __Minimum__ thirty millions (Six million Jews +twenty four million slavs) Killed by the Germans. In\ the second world war. Thus: $Z_C - Z_F = \Delta/\sigma$

$0\int^\infty e^{-D_c.t} F(t).dt - 0\int^\infty e^{-D_f.t} F(t).dt = \Delta/7$

Now if we allow ($\Delta/7$) = Constant of the integration we can integrate both sides thus:

$$e^{-D_c.t} F(t).dt = e^{-D_f.t} F(D_f).dD_f$$

$(1/D_c)e^{-200.D_c} = (1/t_f)e^{-6.D_f}$

$1/D_c . e^{-200.D_c} = (1/6)e^{-6.\times 30}$

$6/D_c =. e^{+(t.D)c} e^{-6.\times 30} = e^{+(t.D)c - 180}$

$\text{Log} 6 - \text{Log} D_c = +(t.D)c - 180$

Therfore the __Total__ __Destructiion__ by the colonialst:

$(t.D)_{\text{Colonialists}} = 180 - \text{Log} D_c + \text{Log} 6 \approx \mathit{180}.$

As you see The __Destructivity__ (D_c) Grows __Asymptotically__ A nd rapidly Lethal. So we have now __Successfully__ Calculated (t.Dc) The __Totoal Destruction By Colonialist.__

Thus by following similar __Procedure__ and __Criterion__ we can obtain any required elements of any two __Zones__ (Chemical? Biological? Geopolitical?etc.) by plugging the __Most__ reliable and __Rekevant__ numbers available.

The Constant Of Destructivity And The Criminal Genes.

" Also we are now in position to summarize what been Discussed in my book (<u>Forbidden Knowledge For The People</u>) Treated under title <u>Criminal Genes</u>(Parts1-72"

"In the previous DIAGRAM While the HEIGHTS of the two areas (YAxis.) Represent total number of candidate orprovisionalcriminal genes !the width of the areas (X-Axis) Represent the number of these MUTATED in to permeant criminal genes " **The Red Monk**

1-Time!Mutations And The Criminal Genes.

Its scientific fact that genes (Any genes)Need time ! Lots of time to mutate. Therefore we can see from the above diagram(Graph A1) How the <u>Time Line</u> on the X-Axis For the colonialists is represented by X(abcd)While that for the Fascist by X(ABCD)Clearly X(ab) >>X(AB) the ratio of the criminal genes(In Percentage) <u>Generated By</u> Fascists to those Criminal Genes)In Percentage) <u>Generated By</u> colonialists =(AB/(ab)

1-   **Some Proofs:**

This book had offered numerous proofs with concrete cases some of it proving how <u>Even</u> the Enemies of Germany concede that the German people are the most honest (least Corrupt)*of all Europeans in their daily dealings with each other's and this is because the crimes committed by the Nazis as atrocious as it were there was just <u>Not Enough Time To Mutate</u> genetically when compared with centuries of British colonialism. While we see inside colonialist nations they can no longer tell their <u>Engineering From Their EnglishVenom</u>? Or their <u>Security From Homosexuality</u> ! Or Their <u>Politicians And Royals From Pedophiles</u> ?All are Drowning in cesspool of corruption Deception!Bigotry and hypocrisy not been known since the days of the pharaohs!*

(See Pages----Total Inversions Parts 1,2,3,etc) ?

(See Pages----Role Inversions Parts 1-6)?

2- The Speed Of Spreading Criminal Genes.
"Like all other genes criminal genes spread in Geometrical Not Arithmetical Progressions! "

<u>Criminal Genes</u> *Can spread much faster therefore :*
I hold very pessimistic view of world that eventually Will be overwhelmingly run by these <u>Criminal Genes</u>.
<u>Again You Do Not Need To Believe Anyone</u> just look at the fact how nations virtually had no <u>Organized</u> Crime To speak of suchas Russia or Arabia nowadays have the worst kind of(Mafia-s) !

London the <u>Capital Of Colonialism</u> itself now <u>The Murder Capital Of The World</u> with kids knifing each other's to death literally on <u>Hourly Bases</u> While the government not only turningblind eye but discretely encourage it!(Deemed to be an effective way to reduce the population of (Wogs)In England).

Inside Many <u>South And Central American</u> Countries who have had close links with colonists : Murders are committed <u>Again ByThe Hour</u> Worst still accepted as the norm ! Nothing unusual about it!!Keeping in mind we are just talking about the crime ofmurder let alone other forms of crime.

3- The Solution?

We cannot stop the spread of these <u>Criminal Genes</u> but we certainly *can slow it down by one solution but unfortunately this solution may be sadly* <u>Misunderstood</u> *as racist because it requires not to have any physical <u>(Sexual)</u>Contacts with any people who had <u>Prolonged</u> Colonialist history such ass the E.S.P. To avoid Reproducing these criminal genes inside other people with <u>Relatively</u> Lower criminal genes (In <u>Percentages</u> per DNA and <u>Frequency</u> strength per capita.)*

<u>*But* **Is** *This Solution Racist* **?its** *Not*</u> *:*

Because it's the only <u>Practical</u> way to prevent this planet from reaching the <u>Final Apocalypse</u> !A world of crime!!Trust me we are not far from it as my book had Already proved in concrete terms how <u>Even</u> the police in <u>England And Wales</u> or the <u>U.S.A.</u> Are Committing More Crimes Than The Criminals Themselves? Simply because criminal genes do not jump out of the skin just because the person adorn police uniform. Period.

"Do Not Expect the criminal genes to jump out of the skin just because the body jumps in to police uniform."

The Red Monk.

"Any so called (Intelligence Services) Like that of the British which thrive itself on driving foreign students to homosexuality or madness : Not worthy Even of its name."

The Red Monk

------- 4- --- **The L.D.G**

But do not get this wrong : Children of colonialists like E.S.P. Can be as innocent as the rest and this is because <u>Criminal Genes</u> Belong to <u>Group Of Genes</u> what I termed scientifically LDG (<u>Latent Destructive Genes</u>) Such as those <u>Responsible For Ageing</u> !

This group of <u>Destructive Genes</u> Normally activated after reaching the <u>Age Of Forty</u> in fact many religions had discovered this fact empirically (By Practice) therefore the age of forty became known to them as the <u>Age Of Wisdom</u> Because the individual needed real wisdom after that age to suppress the urges and criminal tendencies driven by these criminal genes.

(See Pages----Killer Race Parts 1-27) ?

(See Pages----What Is Genetic Mutations)?

The Continuity Of The Constant Of Destructivity,

"Countless millions killed by the British in Asia! Africa!! And the Middle East All had been justified in English Language! By English Logic !Logistics !!And Venom."

The most striking result is how the Ratio of the Buvo to the Euvo (as defined on page ---)?Remain constant even at Peace Time !This can be verified by taking the BUVO Now as the number of overseas students killed or disabled inside Britain since the second world war divided by the population of Britain while the EUVO Similarly measured by taking the number of foreign students killed or disabled inside Europe divided by the population of Europe! ? ! You will find the Ratio Of The Two Rato-S(Plr) Staying constant through out war and peace ! ? !

BOTTOM LINE.

"$(D \times T = T \times D)$ Is a <u>Mathematical Philosophical Field</u> where <u>Causality And Effects</u> Exchange places simultaneously ."

"The most amazing application of my equation:

$(D \times T = T \times D)$ When both D & T Exchange their units"

(See page---Introducing The Heterogeneous Units Of Measurements) ?

"Let us pray my theory is wrong! Otherwise it can only mean one thing: That the Nazis did not reach the figure that was struck by the colonialists? There fore we need to embrace ourselves for the return of the Nazis (Not necessarily Germans) To make up the difference ! To satisfy the constant of the equation."

"Oh! Sinners of the world? This is a call from the biggest sinner of them all: Let us all come clean? Clear your conscience? And ask yourself : Could the Nazis ever match these figures?! **The Red Monk.**

(See Pages----Killer Race Parts 1-27) ?
(See Pages----What Is Genetic Mutations)?

If you happened not to like equations just do <u>Simple Totals?</u>

The number of people the British killed in India alone was <u>officially</u> recognised to be fifty five millions ! ? !

Not to mention <u>Miscellaneous</u> like:

Iraq (Pop-17 millions)lost one million killed and many more dislocated!

Palestine (pop-one million) lost half million and much more dislocated forever,

And the list believe you me is endless from Sudan to Afghanistan.

Now clear your conscience?

Come clean?

And ask yourself:

Can the Nazis ever mach these figures? ! ?

<u>And figures do not lie</u>.

Comparison between (A & B)Any two <u>Entities</u> that are <u>Possibly Nearest</u>than any other two in <u>Size</u> !<u>Mode Of Development</u> ! <u>Religion</u> ! Even <u>Racial</u> (Non Latin)etc.

SYMMETRICAL examples on PROTRACTING TORTURE.

"This Japanese <u>Most Favoured</u> method of torture canact as <u>Ideal</u> definition for <u>Protracted Torture</u> " **Typical example is that practised by <u>Another Island Race</u>? The <u>Colonialist Japanese</u> !Their infamous method of tying the P.O.W. (Prisoner Of War)flat on the ground under a tap water that was only dripping!**

The drops ! will not kill the victim but after few hours or even days you can imagine the rest?

In fact looked at individually the torture may even seem to be trivial !However the picture as a whole convey the most destructive form of torture to both the victim as well as the torturer ! ? ! It's difficult to identify and even more difficult to follow because by definition <u>Protracted Torture</u> spread over lengthy heathen periods and well beyond what human nature can comprehend let alone believe ! ? !

BOTTOM LINE.

Tortures often practiced for no other reason than that of the Japanese or the British they simply have deep seated <u>(Genetic)</u> disrespect for all foreigners in war time or in peace time ! Often just scenting vulnerability in the targeted individual is enough reason?

<u>Protraction</u> seems to be an <u>Indigenous Creature</u> of all <u>island</u> races

In this case [$\delta \theta$ =is represented by the drop of water.].

@ θc---the victim cracks up. And the process becomes more <u>Physical</u> than mental.

@ θ -- the victims kill themselves.

WHERE $\int_c^k d\theta$ - $\int_0^c d\theta = \theta^r$. and r = f(t)

Graph Seven.

THE WORD "BY" WAS DERIVED DIRECTLY FROM EQUATIONS [1+2+3+4] WHILE THE WORD "ON" WAS A CONSEQUENCE OF EQUATION-5.

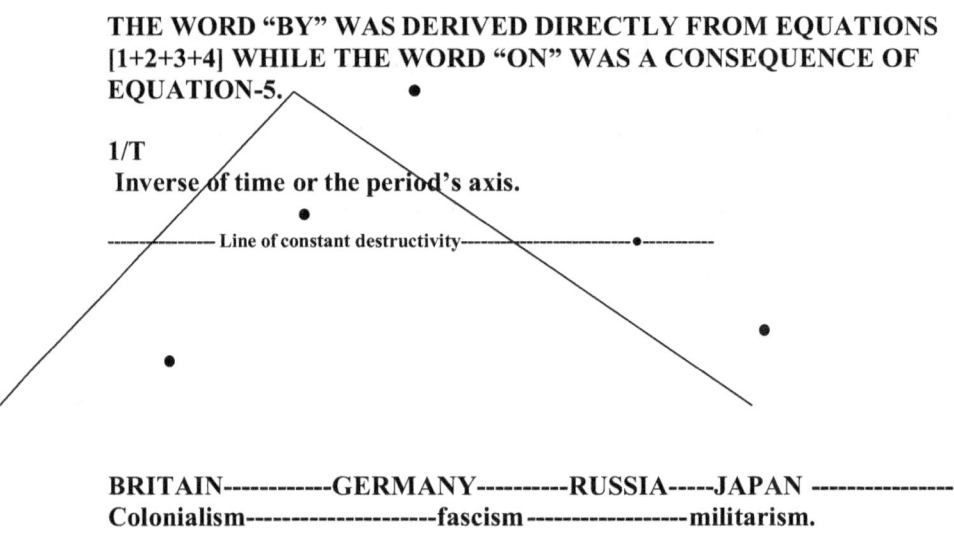

1/T
Inverse of time or the period's axis.

------------- Line of constant destructivity------------●----------

BRITAIN------------GERMANY----------RUSSIA-----JAPAN ---------------→
Colonialism--------------------fascism------------------militarism.

And one practical application of the above formalism that: we can actually calculate the demise (time/amount) of any colonialist by <u>plugging in</u> the values of what we already know of the shock fascist. And how true and prophetic the equations written in 1995 proved to be?

When you study state of the colonialist now days?!

I don't want to go to details that may <u>Remotely</u> benefits the enemies of mankind(the Brits& Jews)but it does not require much intelligence to see how they are certainly the shadows of their previous selves ? !

Industrially? Economically? Politically???

Technologically????Even biologically?????

Not only that but <u>Revenge &Retributions</u> will be inevitable?

They will be hunted down by the <u>Forces Of History</u>!

Inside their own homes for their past <u>Crimes</u> Be it against helpless individuals or genocides alike !

As it was amply demonstrated in no uncertain terms by the <u>Precursor</u> of "<u>Things To Come</u>" by the events of 7/7.

(See Page-----------------Graph Eleven)?

The Politics Of Right Or Left And Two Dimensional Analysis.

If the properties assigned to the <u>Conventionally</u> Accepted notions of right and left are represented by the <u>X-Axis</u> alone (The horizontal line below)?

For example those who are for working class liberation are delegated to the (<u>Left</u>)While those who disapprove of homosexuality and gay marriages are considered to be of the (<u>Right</u>)!!But : Where can we place some one (P-For person) ?Who stands for both i.e. for working class or women liberation as well as for family values opposing homosexuality and is against gay marriages.? Clearly The answer cannot be resolved by one dimensional (The X- Axis)/!What is required is another dimension of (Y-Axis):Where we can assign all the positions taken on <u>Economic</u> questions to this (Y- Axis) !And those taken on <u>Social</u> questions to the (X- Axis)

.For example those who are for the working classes(The Progressives)Can be represented by the upper half of the (Y –Axis)! While those of conservative economic views (The Regressive)Are represented by the lower half of the (Y- Axis.).Hence we can find our person (P) Mentioned above who stands for both the <u>Economic</u> liberation of women or working classes as well as standing for the <u>Social</u> Family values against homosexuality or gay marriages in the first quarter (Upper Right) Represented by the point (P) Below: And so forth!

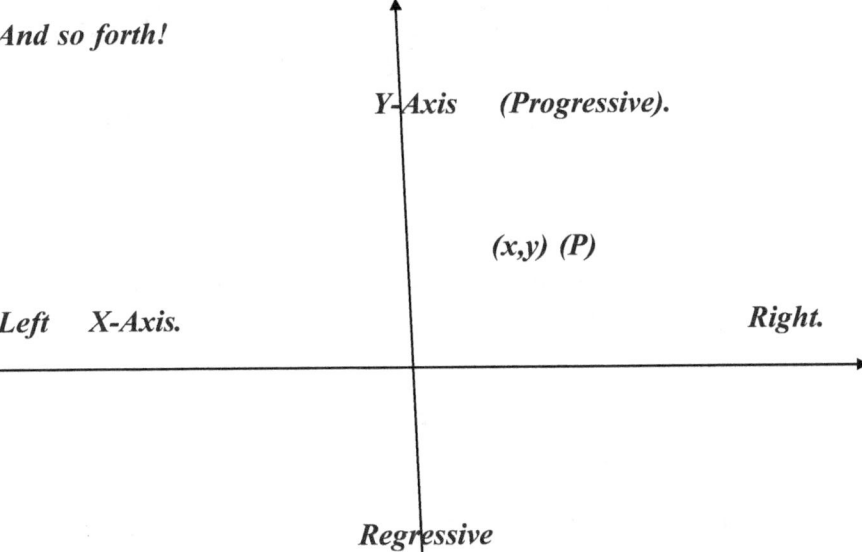

"Since the inception of <u>Militant</u> Zionism at the beginning of the twentieth century :
Political dimensions had become increasingly <u>Ambiguous</u> !
With Zionists acting as <u>Confusion Agents</u> Running like rats inside their tunnels !Inside some secret conduits between right and left so much so that inside the USA itself the far right and Zionists are becoming one and the same !
While in Europe one can no longer tell the difference between their left or Zionists.
In all of this Zionists have one and only powerful tool: And it's not the ideology or any common values shared with rest of mankind because they have none !But it can be summed up with one single word (Money).
How long will this (Power By Gild)Will last is another matter ?
But judging by the past :It will not be long before it will be coming down on their unworthy heads and Their tendency for sodomy."

The Self Destructivity Of Man And The Constants Of Destructivity.

"Most of these equations are just exercises which may not even be fully accurate yet but they prepare us for the real stuff located inside the list of equations at the end of the summary."

$$D_1/P_1 = \text{Constant} = D_2/P_2$$

Or:

$$\text{Any Issue } [D_1 \times T_2]^{d2}{}_{d1} = [D_2 \times T_1]^{t2}{}_{t1}$$

Or Simply: $D \times T = T \times D$

Subscript (1) *Denoting the values for protracting fascists (e.g. The colonialists).*

Subscript (2) *Denoting those values for shock or non protracting fascist e.g. Nazis.*

D= *Is the total amount of destruction per period (P). This can be measured by the number of people killed ? Maimed or their lives are ruined? Or : Number of <u>Square Metres</u> of building destroyed ? Or number of <u>Square Kilometers</u> of crops ruined.*

T= *Units of time inside Real time window e.g.:*
Such window for colonialists like the British can <u>Stretch</u> From the <u>Nineteenth Century</u> Where British soldiers played polo with the heads of Australian aborigines after severing their heads for no other purpose than this **(See Page----English Polo)** *?*
To the <u>Twentieth Century</u> Where British troops literally machined gun hundreds of <u>Unarmed</u> civilians to death at Amstar –India - 1919. To the bloody Sunday when British soldiers killed tens of <u>Unarmed</u> peaceful demonstrators in northern Ireland in the early seventies . To the <u>Twenty First Century</u> When the British were systematical goring(Carving out) The eyes again of unarmed civilians during the Iraq war of 2003-2013.

While such period for the non protracting fascists will be <u>Confined</u> to the narrow window of <u>Six Years</u> of the second world war during which the Nazis gassed six million Jews.

BOTTOM LINE.

"$(D \times T = T \times D)$ Is a <u>Mathematical Philosophical Field</u> where <u>Causality And Effects</u> Exchange places simultaneously."

$$D_1 > D_2$$
$$T_1 > T_2$$

And for absolue values :

$$T_1/ D_1 > T_2/ D_2$$

Comparison between (A & B) Any two <u>Entities</u> that are <u>Possibly Nearest</u> than any other two in <u>Size</u> !<u>Mode Of Development</u> ! <u>Religion</u> ! Even <u>Racial</u> (Non Latin) etc.

(See Page----- New Physics) ?

How Much More Can We Read In To The Equation Of :

D X T = T X D

"There are more to this equation than meets the eyes! However we left it as exercise to readers to help them <u>Focus</u> on this equation? Here is few hints"

1- *It means destructivity is constant.*

2- *All parties plugged in to the equation will strive to satisfy the constant numerically i.e. the colonialists will not rest? Cannot rest until they kill same as the Nazis! Same as the Stalinists etc! i.e. there is an involuntary mechanism .*

3- *Germans just do not have the stomach as the British to maintain such atrocities in Long Protracting <u>Cool Calculated Callous</u> Fashion <u>Never Ending Agonies</u> in time of peace! (see page on the three C`s).It's just not in the German character and outside their abilities to sustain a British-like campaigns of venom forever.*

4-   Hence the only <u>real difference is the timing ? The period these atrocities against humanity take ? That is all.</u>ie;
Again scientifically speaking the British practice "things" like mass killing in <u>Times Of Peace</u>(e.g. that of overseas students) to make up for what the Nazis practiced in the <u>Short Years Of The War</u>.

5-   If you study closely the psychology of the <u>average</u> Briton (Who have not been told yet that his Britain have <u>ALREADY</u> killed more people than the Nazis ever did)You will see him having a grudging feeling of being short changed that they(the Brits)did not kill as many as the Germans !

<u>Therefore They Had Lots Of Catching Up To Do</u> !

Indicating the existence of An <u>involuntary Motivating Force</u> inside him !A <u>Concrete Scientific Evidence</u> that this average Briton <u>Carry A Charge</u> inside driving him consciously or otherwise to satisfy the constant of the equation numerically i.e. by striving to kill as many as their next door (The Joneses)Had killed ! ? !

6- More Interpretations:

"Whether it's by <u>Gun</u> or by the <u>Poison Bottle</u> The <u>Criminal Gene</u> that runs inside every E.S.P(And the –P- Now for Person) Does not care !And waiting there to manifest itself one way or another"

Once upon time at a bus stop I entered in to casual conversation in Arabic with an extraordinarily good looking Algerian. Of white complexions and tall dimensions. While waiting for the bus we shared our impressions of the world telling me that he went to the USA to study English but found it very unsafe due to the gun violence there. Hence he came to Britain ! Yet again he still feel unsafe here !! I do not know what its?? But there is something not right about these people too ?Though I cannot put my finger on it! <u>I Felt Safer In France Although The French Hate Us(Meaning The Algerians).Said He</u>.

Then I asked him if he had ever as a kid watched Tarzan movies? Where the extremely noisy forests crowded with wild animals

Suddenly goes silent! Falling in to deafening silence ?

Reason being ? Though these animals had no radars or cellular technology but instinctively could feel approaching dangers including thunder and lightening. Nature has endowed all of us by this <u>Instinctive Radar</u> !By this <u>Sixth Sense</u> !To sense forthcoming dangers. You see what this Algerian had felt is the two sides of the coin? Two manifestations or expressions for the <u>Constant Of Destructivity</u> described by the given equation : (D XT = T X D.)! Where the <u>Open</u> gun violence inside the USA was <u>Replaced</u> by the <u>Clandestine</u> destructivity of poison bottles and protraction of the U.K Practised <u>Even</u> by the British state itself!

Or vice versa that is to say: The <u>Sneaky Clandestine</u> British ways of destructivity <u>By Stealth</u> ? <u>By Protractions</u> ?<u>By Poison Bottles</u> !etc . Have <u>Re Emerged</u> on the other side of the Atlantic at the USA with <u>Visible</u> gun violence etc. The two sides of the <u>Same Constant</u> In the above equation.

Another Significant Application For My Equation :
D X T = T X D.

Translation:

"Fascism is <u>Self Destructive</u> Evil !Colonialism is <u>Self Perpetuating</u> Evil. "

"Due to the short outbursts of <u>Shock Fascists</u> Crimes(Maximum six years in the case of the Nazis) The <u>Criminal Gene</u> was not given enough time to <u>Mutate</u> Throughout the population as it was the case with the <u>Protracting</u> Fascists (<u>Centuries of colonialism</u>) e.g. the British)And here <u>Once More</u> we have discovered another <u>Significant Application</u> for my equation. Appearing in many shapes and forms inside this book ? Rest is left as an exercise for the reader ?"

"Hence you note the people of Germany are far morehonest than any other Europeans <u>By Any Standard</u>.""Whilst in Germany you still see racist thugs on theirstreets! In Britain these racist thugs are <u>Firmly And Cozily Hidden</u> Inside the power points of the Britishstate itself."<u>**For More Application Of My Equation See**</u>: *Introducing The Heterogeneous Units Of Measurements.*

The Scientific Relationship Between Fascism And Colonialism.

$$D \times T = T \times D.$$

"The above equation appearing in various shapes and forms at several stations inside this book together with the help of <u>The Scientific (Statistical) Tables Of Destructivity.(Part –Two).</u> Should have made us all acquainted with its deeper meanings ?Here we take another look at this equation relating fascism to colonialism in concrete scientific terms:"

"It's no coincident that the <u>Nazi Salute</u> was identical to that of the <u>Romans</u>."

"It's no coincident that the royal family of Britain had been photographed performing the <u>Nazi Salute</u> ."

Most people are shocked by the latest(July-2015) photographs which was _Accidently_ Released !

Proving how deep and direct had been the connections between the royal family of Britain and the Nazis of Germany ??

Some royals were _Even_ Blood relatives with senior gestapo officers !

Others were dining and winning with Hitler himself !!

Other photographs showed them wearing Nazi uniforms and performing the _Nazi Salute_ !!!

These latest revelation may have had shocked most people except those who had fully understood both the _Implications And Applications_ of the above equation relating fascism to colonializing in _Concrete Scientific_ terms.

You see like the romans like the Nazis the _British Royal Family Inner Thinking_ Proceed from the premise of the _Absolute_ !

That Is The Absolute Right To Absolute Superiority!!

They interpret their <u>Absolute Privileges</u> a divine right that must have descended upon them from god himself by virtue of the fact they remain unchallenged(the fact they had installed themselves at several points of history by the sword and this right was not descended upon them from god but had descended upon human skeletons ! ? !)

Such facts had been conveniently misled inside the countless pages of history !

In the case of the Nazis they assumed the <u>Absolute Superiority</u> of the Aryan race (e.g. by virtue of the fact they made better cars)! ? !

Similarly no colonialist can colonize anything without the brutality of proceeding from similar <u>Platforms Of Arrogances</u> :

Convincing themselves one way or another of the absolute unquestionable right to destroy! I said (<u>To Destroy</u>)Because destruction is inevitable inside any such criminal enterprise called colonialism !

Since the <u>Colonialized</u> Like all victims of crime will not be greeting the invaders with flowers. Moreover they all destined to fail ! Like the Nazis they all are doomed .

<u>*As To What Is Keeping The Royal Family Of Britain ?*</u>

<u>*Is The Duality Of British Society !*</u>

<u>*But This Cannot Go On For Ever!!*</u>

<u>*At Some Point In The Future This Duality Will Split Britain Itself In To Pieces And No One Can Do Anything About It!*</u>

<u>*Not Even Their American Dogs*</u> *!!!*

(See Pages--------------Duality Parts 1,2,3,4,5,6,7 ,Etc) ?

HATE MACHINES(PLR).

"In socio political affairs as in the physical world there is a concrete relationship ((The Longer (By Stealth)Is Practised ! The Lesser Will Be The Feedback !The Bigger Is The Lag (Phase)Angle)."   **The Red Monk.**

"To kill and maim such huge numbers of people! The Nazi of Germany (Anglo-Saxons) and the colonialists of Britain (Anglo-Saxons) needed not only bullets and helicopters but also <u>colossal Hate Machines</u> that begins at their <u>Kinder Garden</u> schools"

Now we are in a position to calculate the amount of pure venom by the A.U.V.(The Anglo-Saxons Unit of Venom).

A.U.V. =*The total number of people killed by the said (corresponding) Anglo-Saxon country [÷] {Divided} By the total population of that country.*

Probabilty !Triangular Inequality :The Contradictions.

"There are contradictions in the Mathematical Concept of probability."

"Unless we can guarantee a result after certain number of throws(e.g. Of the Dice)The whole concept of probability remains problematic." **The Red Monk.**

Unless it can be absolutely guaranteed that we get an <u>Event</u> If we throw f(n)-times for probability of(P_n) For such event(S) Then the very concept of probability itself must be flawed! <u>In Fact There Is No Guarantee If We Throw Million Times.</u>

Consider a particle at the point (A) Intending to reach the point (C) There are two ways for this :

The straight path (AC) or the longer path (ABC)!!

Logically speaking the particle will take the shortest path and we say the probability for this:

(P_{AC}) Is higher than that for the longer path (P_{ABC}).

I.e. If we allow (AC) To represent the first probability and (ABC) The second then the triangle inequality is <u>Violated</u>!

<u>On the other hand</u> *The only alternative left is for the inverses of these length to represent these two probabilities. i.e. $1/AC \geq 1/ABC.$ $1/(AC)^2 \geq 1/(ABC)^2.$*

But:

This too violate the triangle inequality at <u>Right Angles</u> Since :

$(AC)^2 + (AB)^2 \geq (AC)^2.$ *I.e.   Again contradictory*

BOTTOM LINE.

Contradictory ?

Or Meaningless Invention ? ?

Its on upon this <u>Opposing Principle</u> that <u>Quantum Theory</u> should be <u>Reconstructed</u> ! Instead of basing it on extended concepts of classical probability.

At The Point Of Separation.

Imagine there is an ensemble of one hundred and twenty identical macroscopic objects e.g. balls ?
They are governed by one single probability (Distribution)Curve? Then if we are to separate them in to two batches of sixty balls each ? There will be two curves instead of one overlapping each other's at the <u>Moment Of Separation</u> The question now :
What happens if we to simultaneously separate this ensemble in to three batches of forty balls each ??

Do We Obtain Three Dimensional Curves?

And If So What Does It Mean??

Exercise Seven:

How can we relate this hypothesis to Graphs numbers : One? Two ??And Three???

Examples and Applications.

The Multiplier. $\lambda_i = m_j - n_k$

"(λ_i) The Multiplier Is the <u>Mathematical Tunnel</u> connecting our <u>Economical</u> activities with the <u>Socio-Political</u> To the very <u>Physical</u> World of physics and geometry."$\sum$

Most of the methods and equations described in this book were borrowed from the methods of mathematical physics when solutions sometimes are pulled out of the blue so to speak ! Then tested to see if it was the correct solution! Its called(Back tracking).

Similarly we related economical classes ! Their behaviour !! And the outcome of their behaving !!!By assigning values such as $(n_1 . n_2 ? n_3 ?$ et!$)$ These values may seem to be too arbitrary! Unfair or Unfounded !until we plug in some economical values in to the equations scattered through out this book such as the one above to see what happens?

KARL MARX ? L T V ?and Reality.

LTV=Labour Theory of Value.

R=Rate Of Profit. S=Surplus value K= Fixed(Not Constant) Capital.

In Chapters Nine And Ten (Das Kapital)Karl Marx Struggled With The Value –Price-Labour Relationships introducing as first approximation $R(r_i ? r_j ? r_k ?)=S/K$? Eventually he left the issue unresolved and quite rightly so because its like rocket science too many variables and Never Enough Integral Equations. So did Adam Smith recognising the complexity of the issues involved (Both Marx and Smith Were not mathematicians but Maynard Keynes was). They all seems to have had underestimated the complexity of the calculations involved !

For where does (R) goes ?

The flow of money locally or globally ? etc? etc ??

Denoted As Follow:

How was it distributed?

(i , j , k,... etc. Respectively).

How its invested?

($i+j+k,+...$etc. Subsequently).

How is it Reinvested ?

($I \times j \times k \times$...etc. Consequently).

They are all complicated series(es) Each with life of its own! All concurrent ! And none congruent !!

However in this book you will come across many equations offering simple but clearer guidance to explore any uncharted opportunities offered as in My second approximation $e_{i,j,k}$

$$i \quad j \quad k \quad \lambda_{i,j,k.} \quad i,j,k. \quad i,j,k.$$

Where **($\lambda_{i,j,k.}$)As** **defined** **above.**

Exercise.

Its left up to the reader's insight !expertise! And versatility to plug whatever economical or social numbers in to the equations offered by this book :

1-How can we relate (r-the rate of profit)to Maxwell - Boltzmann distribution? And to the equations inside this book such as the one above ?

2-Is this the reason ? Why we left (m&n)Deliberately undefined as mass or whatsoever? The <u>Multiplier</u> =m - n <u>(In this particular

$$\lambda_{ijk}$$

case : i,j,k. Are time ordered.)</u>

If you plug the values in to the appropriate equation from thisbook would it be nearer to reality than the extremely complex equations familiar to economists relating quantities such as: <u>Price Of Production To Value To Labour</u> etc?etc??

4-

How can you fit the global tendencies for R(Rate of Profits)In to the General picture of:

Ψ Economical Equilibrium. *Stated at the beginning of this chapter?*

(See ?Hasok Chang Inventing Temperature : Measurement And Scientific Progress -Oxford - 2004.) ?Also See:

(Francoise Quesnajin (1759) And The French EconomistsKnown As The Physiocrats.) ??

Also See:

(Forebenius Mathematical Models.) ? ? ?

Power And Production.

"Perhaps its purely coincidental that both <u>Power</u> and <u>Production</u> starting with the letter (P) In <u>Latin</u> ?? Perhaps its another hint from metaphysics for us to take closer look at this irony ? At the relationship existing between the two ?? Most appropriately after the above discussion on positive and negative professions ??"

These two are more related physically and metaphysically than has been so far recognized !

Production is simply dictated by the need of the masses to survive!

Power is dictated by the need to organize production !!

Power feed or leeches on production. yet paradoxically it hinders production by its own logic that have developed and has taken life of its own i.e. the logic of power gradually but surely will be diverging from what is necessary for production ! ? !

Therefore production aims to abolish or at least limit power !

*But power by now and <u>**By Definition**</u> Resist such tendencies again by definition with all full powers available to its own !*

The paradox is both need each others but only in <u>Certain Ratios</u> Which is a must for healthy coexistence ! However once this <u>Ratio</u> Is violated then both try to eliminate each others by fighting each others ! The process is that of conflict ! Adjustments and ! Evolution created by the need to each other. This ratio is expressed by the letters (m & n)Inside the first arithmetic ,

(See Page---The First Arithmetic On Revolution.) ?

BOTTOM LINE.

"Power without production is Paper Tiger"

The Red Monk.

"----Because Real Life is not about the profits made by few Jewish Bankers! Or few rich homosexuals leeching on the Derivatives Of paper money !! Real life is about everyone having the right to get up in the morning feeling needed ! Wanted !! And going to work in real jobs ? Then returning home feeling he had earned his living by fiat not charity .

A sense of satisfaction snatched in modern times from the average homosapiens by Jewish bankers and the leeching parasites of state homosexuals."

By Definition:

1- Every thing in the universe is quantifiable! The question is how ?

2- There are no <u>Massless</u> particles ! If we perceive or conceive it to exist then it must have mass(Including Photons.)!!The problem is that of measurements?

3- No speed in the universe is constant ! The <u>Answer</u> is in the question : Why?

4- Every geometrical line in the universe that is not mathematically trivial e.g. ($c^2 = a^2 + b^2$) Represent concrete physical existence! The issue is that of cutting and gluing the line simultaneously ? If the diagonal of rectangle !Triangle etc. can be related algebraically to others e.g. Its sides then every Such line potentially has hidden agenda waiting to be discovered.

5- There are (At Least)As many life forms as the number of elements in the Mendeleev table <u>If Not More</u> !Except that no physical contacts are allowed ?

On Information Theory And The Limits Of Hypocrisy:
(Part-One)

"The secret never found in the lie itself ! But inside: Why is the lie ?" **The Red Monk.**

"Hypocrisy is misinformation! And misinformation is hypocrisy." **The Red Monk.**

"You see it's not just about ethics: Mostly about how hypocrisy allow the creeping of dangerously skewed cognitive process totally at odd with reality and hindering the ability to analyze reality correctly ."

"Whenever I read German or French newspaper I am looking for the news! When I read British newspaper I am guided by one and only one quest: Why are they lying ?This way I manage turning their <u>Sheets Of Lies And Deceptions</u> (Including self-deception)In to an <u>Encyclopedias Of Real Information!</u> But it requires intelligent readings and some <u>Direct</u> experience with the British art of <u>Double Deception</u>. "

"A nation like the Brits is <u>Doomed</u> by its own hypocrisy : You see hypocrisy is not just about ethics !In practice it create <u>Dangerous Misundestandings</u>! <u>Miscalculations ! And Irreversible Lag (Phase)Angles.</u> Its when men short change themselves by getting too clever."

Practicals :   On The Limits Of Hypocrisy:

"To comprehend the limits of hypocrisy in tangible practical terms all you need is to examine the Britishcase :"

"The consequences of any <u>Invisible</u> Major or minor crimes committed by the colonialists :Sooner or later becomes <u>Visible</u> But by then it will be very much <u>Irreversible</u>."   **The Red Monk.**

"My own <u>Practical</u> Experiences with the world that is both the written or unwritten world are telling me that you can sit on anything but not for long !And you can pretend everything but not for ever !!So why the delay ?Why operating at this vast <u>Lag (Phase)Angle</u> Between illusions and the reality of their own size and resources? It does no one any good whatsoever???"

Here the <u>Lag (Phase)Angle</u> <u>Between British Ambitions And The Reality Of their Size Or Resources</u> is as massive as black hole !And its doing them more harm than any good. But the Brits do not know or do not want to know that because they are hypocrites !Caught inside make belief world of dancing in the rain. Drowning in universal hypocrisy spreading from tampering with the holly grail (<u>Police Log Book See Page---</u>) !To fiddling most essential government statistics !To doctoring the daily news !!!!Perfect practical illustration :While the role of the press in most democracies to expose corruptions you will find the role of the press inside this British <u>Pseudo</u> Democracy Is to cover-up corruptions as we had seen elsewhere in this book how the British press cheekily and unashamedly calling for the police to apologize for investigating crimes of sexually assaulting boys or in some cases even murdering them just because these ghastly crimes were committed by high ranking army and political personnel ! ?

(See Page---Pseudo Democracy) ?

When The Differences Between The State And Institutionalized Crime Reaches The Vanishing Point.

"$N =\sum_{0}^{n}(n/2)\pi$.Hence: L- N.I =0(See Page-Angle of lie)"

So if Britain is **Doomed** .How did they manage so far ?? Simple: **By Compensation** !By compensating their short falls with **Excessive Brutalities** !**Untold Cruelties** Some had been described by In this book under the title of **Written Off** Ranging from sending their own (Commoners)All the way to Australia in chains to die with dysentery! To the more recent poisoning or burning to death of (Wogs) Inside Britain itself!(See Silent Genocides)?Also by **Protraction** and second to none system of (**Doing It By Stealth**) !

Briefly by <u>Institutionalizing Crime As Source Of Large Part Of Their National Income</u> To the extent that decent behavioral living is deliberately criminalized!

<u>It's When The Differences Between The State And Institutionalized Crime Reaches The Vanishing Point.</u>

<u>However</u> with the advent of I.T This (<u>By Stealth</u>) Factor will be increasingly debilitated if not diminishing altogether and all these <u>British Peculiarities</u> Will be exposed to the world <u>Systematically</u>! World reaction may be slow patchy and latent but definitely inevitable!

(See Page_____-_____Angle Of Lies)?

(See Page----------------------------Material Dialectics Part-3)

(See Page-------The Truth Index) ? (See Page-----Lag Angle) ?

Machine Calculations:

"Here we provide an alternative to the information theory proposed inside the <u>Fifth Arithmetic</u>!"

To simplify the ideas proposed earlier (The <u>Fifth Arithmetic</u>):Let us say there is machine (M) Where [M f (T).]That recognizes a lie (L)when it does not fit in to the general truth(T)!But what is the <u>General Truth</u>?If we say (<u>All Men Are Idiots</u>) Or (<u>All Idiots Are Men</u>) Or (<u>All</u> <u>Men Are Women</u>)The machine (M)will recognise these nonsensical statements <u>Directly</u> from the <u>Mechanical Position</u> ofthe word (All) !And the machine need not be programmed with anything <u>Qualitatively</u> More complicated than your average computer! Applying similar <u>Logic Gate</u> Only with slight modification .As follows: All Men Are Idiots → Men Are Idiots →Men Idiots →Idiotic Men. etc ?

Let (L-For lie) Be the value in <u>Known Kilobytes</u> of such misinformation.

Let (T)=Truth. In <u>Unknown Kilobytes</u>?

Then:

$M(L) = \Delta L + \Delta T.$

Now we insert these values back in to the machine:

$M(\Delta L + \Delta T.) = (\Delta L)^2 + 2\Delta T).$

And so forth we keep <u>Pumping Back</u> all these values for the machine not only to churn out figures ? But also acting as <u>Filter !</u> Or even a <u>Converter</u> !!

Giving :

$M\ (\Delta L)^2 + 2\Delta T) \rightarrow (\Delta L)^n + nT.$

Or:

$M(L) - \sum_{n}^{0} (T) = (\Delta L)_n$

$M(L) - \sum_{\infty}^{0} (T) = 0.$

Quantum Thresholds.

"Quantum threshold is not a barrier! It's not there until you try to climb it !! These are no ordinary barriers!!! It only erect itself instantaneously the moment you try to climb it and only after certain factors (Conditions) Had accumulated on both sides of the barrier !!!!"

"Although it may be ignited by inequities!
*Revolutions do not happen **Just** because of inequities!*
The world is full of inequities!
And probably shall remain so indefinitely .
Revolutions are all about <u>Energizing</u> !
Rejuvenating the society!!
<u>Again You Donot Need To Believe Anyone</u> :
Just look at the <u>Quantum Surge</u> in discoveries and innovations of the arts and sciences which followed each and every revolution in history ?

Society without successful revolution (like Britain) is a dead society except by name !Revolution is Recharging a battery with the <u>Poles</u> being the following (m&n):" **<u>Three examples already discussed by this book</u>** :

827   147

New Physics. New Philosophy.

"It's well to remember that highly prestigious universities (e.g. Scottish universities)Call their departments of physics instead : Departments of <u>Natural Philosophy</u> Because physics except for the ignorant is the philosophy of nature .

"We all familiar with the <u>Spinning Top!</u> Floating around defying gravity before our very own eyes."

1- Instead of saying:
The airplane flight is due to the **Pressure Differentials**
At the foil (Wings) ! I claim it fly because it cuts the cosmic flux at optimal angle.

2- And that the **Gyroscope** Behave the way it does because it's **Cutting And Gluing** The cosmic flux between itself and its surroundings instantaneously.

3- Instead of the rocket flying due to action and reaction :
This view **Claims** because its moving along the lines of the cosmic flux negotiating its path along not across ! Between the lines of flux.

4- And that the **Earth** Stays in its orbit due to two actions :

a- Cutting the cosmic flux between itself and the universe by **Rotating** around its own axis.

b- Cutting the flux between itself and the sun by **Revolving** around the sun itself.

5-

They say that (Airships) Were unstable disasters due to several factors (e.g. Climate)!

My **Alternative Physics** Claims that they were just not cutting and gluing the cosmic flux fast enough !! i.e. there is **Lag Angles** between the cutting and gluing of the cosmic flux.:

Lag (Phase) Angle = $f[\omega_{Cutting} / \omega_{Gluing}]$ Was just too big resulting in (Hunting) Or Anti- Resonances .

The Philosophy Behind My Alternative Physics.

"If we replace the classical frameworks constructed with velocities by translating all velocities (Vectors) In to frequencies (Scalars)! Then a new world with new conceptsmay emerge:"

In Classical Mechanics This Will Be:

$V = \omega \cdot r$

In Quantum Mechanics:

$Re [e^{i(\omega \cdot t - k \cdot x)}] = \cos(\omega \cdot t - k \cdot x)$

EQUATIONS OF MOTION.

"Every <u>Stable</u> object in the universe microscopic or macroscopic has two frequencies ($\omega_{Cutting}$) Which is cutting the cosmic flux between itself and all surrounding objects ! And (ω_{Gluing}) Which is gluing this cosmic flux <u>After</u> its been cut in time (Δt)."

"The trick is to think of (ω) As pair! One of them is just Before reaching its own final value in time (Δt) "

First Equation Of Motion.

We split $[J\omega]^2$ In to $+\omega = \omega_{Gluing}$ And $-\omega = \omega_{Cutting}$

For one dimensional movement along the X-axis :

$$[\omega_{Cutting} - \omega_{Gluing}\; 0\text{-}J\,]\,\theta = d \quad \omega = K\,dx.$$

$$[\omega_{Cutting} - \omega_{Gluing}\; 0\text{-}J\,]\,\theta \sim d2 \quad \omega = K\,dx.$$

$$[\omega + \omega\,]\,\theta \qquad 2$$

Cutting Gluing 0-J

The **Full** expression on the left I had in mind is more complex than what can be printed here! However to **Present The Bare Idea**. These expressions should Suffice.)

Clearly : k = $1/r_{i,j,k}$.. Where $r_{i,j,k}$ = The radii (Plr).
Similarly: $d\omega = K\,dy$. And $d\omega = K\,dz$. Etc.

Exercise Eight:

If one of these (ω)s Is known as the **Natural Frequency** ? Can you determine the other (ω) From the following constellation of objects :

X,Y,Z
$\Sigma k(dx,dy,dz)$.

For objects :
(O) Placed at the origin? Object (A) Placed at (-x)? (B) Placed at (+x)? (C) Placed at (+y)? (D) placed at (−y) and (E) at +z ? (F) At(−z)? etc ? **Can Any Of This Relate To The S.H.O ? Or S.H.M .??**

Second Equation Of Motion.

"The intersection of two lines is the point.
The intersection of two areas is a line.
The intersection of two volumes is area.
The intersection of higher dimensional objects in the universe is the frequency (Multiplied Off Course By The Relevant Vector ($r_{i,j,k,etc.}$))"   **The Red Monk.**

Let (a^n) And (b^n) Two objects of dimension (n) Where [a > b] ? Their intersection is (b^{n-1}).
Let ;

$$1/\omega_{Cutting} - 1/\omega_{Gluing} = \Delta t.$$

And

$$[\omega_{Cutting} - \omega_{Gluing}]_{\theta - J\theta} = \Delta\omega$$

Then:

$$\Delta\omega \cdot r_{i,j,k,etc} = b^{n-1} / \Delta t.$$

But;

$db^n/db = b^{n-1}$

$(\Delta\omega \cdot \Delta t) \cdot \underline{r}_{i,j,k,etc} = db^n/db$

$-(\omega_{Cutting} - \omega_{Gluing})^2 / \omega_G \omega_C = db^n/db$

After Substituting ! Rearranging !! And Approximating !!!

$[\Delta\omega / \Delta\omega_{G,C}]^2 \approx -(db^n/db)$

Keeping in mind that both ($\Delta\omega$) And (Δt) Are infinitesimal.

<----------------------------> dω ? dt ??db ???

Calculating ω_{Cutting} And ω_{Gluing}.

"If in the following diagram the states Ψcutting and Ψgluing are represented by (AB)and(BC)respectively i.e. Orthogonal ?

Can (BD) Represent the state ΨBD in 3D ??

1- Moment of inertia is a function of its own geometry.

Eery point can be assigned a frequency of oscillation around an arbitrary or actual point (e.g. the centre of the earth) !

2- Each line (AB) Can be represented by either the totalsum of these frequencies $= \sum f_n$!

Or the average [$\sum f_n /n$] !

Or simply the differences $= f_A - f_B$

However once the appropriate method had been determined it should be maintained throughout the geometrical construction.

3- <u>Alternatively</u> Every line has signature frequency that distinguish its Length? Shade??And Colours from all other adjacent lines (Spectra).

If F= Function of $(2 \pi . r)$ And f =frequency representing the corresponding lines in the (Isamic) Diagram below:

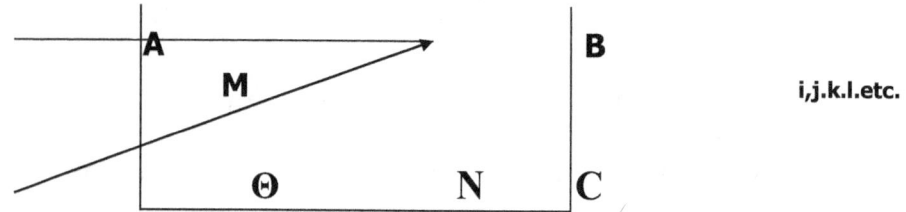

$\omega_{Gluing} =$ The areas $(m + n) = F/2 (f_{ab} f_{ad} + f_{bc} f_{dc})$. And

$\omega_{Cutting} =$ Area $(m - n) = F/2 (f_{ab} f_{ad} - f_{bc} f_{dc})$. For higher dimensions than areas: $(m \pm n) = F/n (f_{ab} f_{ad} f_{ac} \text{----} f_{an}) \pm (f_{bc} f_{dc} f_{ac} f_{nc})$.

$\omega cutting = 0$

At m

$\omega_{Gluing} = 2m = 2n = D n^{D/2} = D m^{D/2}$ (See Graphs 1-9) ?D = Number of dimensions higher than that of area.

Disproving (C).

C = Constant Velocity of light.

$R_{a,b,c,etc.}$ = The Radii from each of these points to the arbitrary or real centre of rotation .

If $V_{a,b,c,etc.}$ = Variable velocities of <u>Sensor Sources</u> Placed at these points a,b,c,etc On the line (DB) Then:

$$f_{a,b,c,etc.} = V_{a,b,c,etc.} / R_{a,b,c,etc.} \quad \text{Or} \quad f_n = V_n / R_n$$

And the average : $\underline{V} = \Sigma f_n / n \times R_n$ (Since R is variable).

Proving (C).

If we now place light sources at these points along the line(DB)Where $V_n = C$ Then: $f_n = C / R_n$

<u>And The Average</u> $\Sigma f_n / n \times R_n = C$

Which is meaningless since all these Points! Lines! Areas etc. Have the same frequencies for the same diagram therefore there is only one value for the velocity and that is the constant (C)Velocity of light .

Geometry With Teeth.

"In this analysis geometrical constructions are treated as physical objects! With frequencies (By painting each line or side with different colours in the Spectrum) !! Density=p (Row) : The number of molecules per unit area in the laminas (Before stretching) etc?etc??Then:"

$$+\delta m = -\delta n = -\delta\theta_m \, \delta r \times \text{unit length} = -\delta\theta_n . K = \delta (1/p)$$

Now do not forget inside ($\psi_{i,j,k,l,etc}$) If i=Represent state of electrical properties e.g. semiconductors ! K=Molecular texture?? etc. then :

$m_i \neq n_i$ And $m_k \neq n_k$ etc?

At: m=nδθ =0.

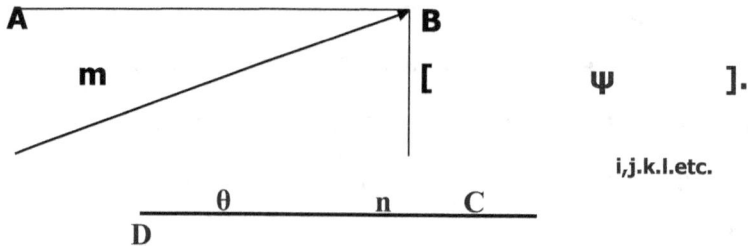

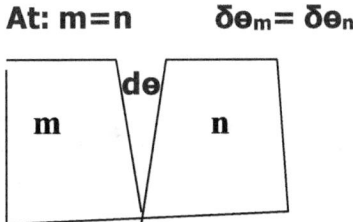

Figure one.

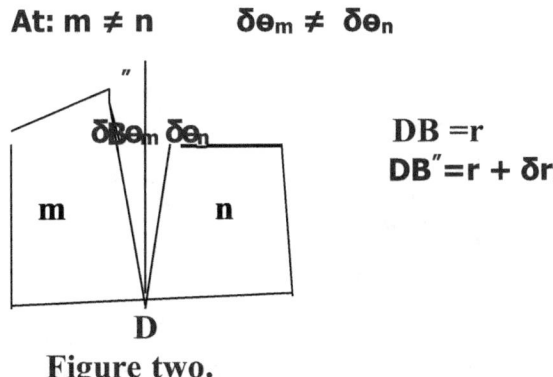

$DB = r$
$DB'' = r + \delta r$

Figure two.

$d\theta/dt = \dot\omega$ $d\theta/dt = \dot\omega$

m   Cutting   n   Gluing

Test:

If in figure one : $d\theta = \delta\theta_m + \delta\theta_n$

And in figure two : $\delta\theta_m = -\delta\theta_n$

Compute: The two ($\_\dot\omega$)s

Probing Higher Dimensins By Physical Frequencies.
Introducing Two Frequencies:

A-   **Mechanical Frequencies (F):**

Spectator /Spectrometer.

Blue   V

              Red

              Radius(R).

The circle above represent the earth with radius (R) Rotating at **Constant** Frequency (F) !At the surface of the earth geometrical construction with one face in the plane of the paper the lamina ABCD (As in the Isamic Diagrams)!Then every single point on this construction should have the same **Constant** Frequency(F).

B-   **Specrtal Frequency(f_n) :**

We paint the line DC With the colour (Red)And DA with (Blue)And so forth with the rest of the construction in different colors !The **Spectrometer** will register frequency (f_{Red})For DC and (f_{Blue}) For DA. And so forth for the rest of the construction(s).

3- Relating Mechanical And Spectral Frequencies.

"Proceeding from the Anzast that frequencies are Scalars! While velocities are <u>Vectors</u> that can be added or subtracted.±" **Essentially by turning the frequencies in to velocities as follows:**

a- We multiply both sides of the equation by unit vector ($\underline{r}$)
B- Then by Wave Length ($\underline{\lambda_n}$) Where n=Number of dimensions.
C- Then removing $\underline{r}$

$F.R.\underline{r} = \Sigma (F \pm f_n).\underline{r}.\lambda_n$ Giving:

$F.R = \Sigma^{n}{}_i (F \pm f_n).\lambda_n$

$F.R - n.F = \Sigma^{n}{}_i (f_n).\lambda_n$

$F = \Sigma^{n}{}_i (f_n).\lambda_n] / (R - n)$ ------------------------- Equation-One.

4- How To Neutralize The Doppler Effect?

By averaging its effect:

$V^2 = [(f_{Red})^2 + (f_{Blue})^2](\lambda_n)^2$
$\quad + [(\lambda_n)^2 + (\lambda_n)^2](f)^2$ Etc. $] / n^2$ Equation-Two.

Now differentiating both sides of **Equation-One** with respect to time! Then substituting for **Equation-Two** We obtains:
O=dv/dt Hence velocity here is constant !
Also giving;

5- $F = \int d/dt \left[\sum^{n}_{i} (f_n) \cdot \lambda_n \right] / (R-n)$.

For example to see what is happening at the 26th Dimension put n=26 And the units will follow?

6- **Alternatively**:

$F.R = \left[\sum^{n}_{i} r_i \cdot \cos(\theta_i) \right] / n = \left[\sum^{n}_{i} f_i \cdot \lambda_i \right] / n$.

$F.R.n = \left[\sum^{n}_{i} r_i \cdot \cos(\theta_i) \right] = \left[\sum^{n}_{i} f_i \cdot \lambda_i \right]$

Where :

r_i = The distant between the point (i) And the centre of mechanical rotation.

θ_i = The angle between (R) And (r_i).

f_i = The spectral frequency recorded for point (i) By the spectrometer.

λ_i = The wave length at that point.

Now differentiate all sides of the above equation w.r.t:

$$d/dt \left[\sum^{n}_{i} r_i \cdot \cos(\theta_i) \right] = d/dt \left[\sum^{n}_{i} f_i \cdot \lambda_i \right].$$

$$0 = \sum^{n}_{i} [dr/dt \cos(\theta_i) - r \cdot \sin(\theta_i)] = \sum^{n} [0 + f_i \, d\lambda/dt].$$

$\sum^{n}_{i} \tan(\theta_i) \cdot r = - r \cdot dr/dt$. And $d\lambda/_i dt = 0$

Giving :

$\lambda = $ Constant .

In the limits :

$2\pi \sim -$(Minus) dr/dt.

Since :

$df/dt = 0$

And since :

$\sum^{n}_{i} \tan(\theta_i) \longrightarrow \sum^{n}_{i} (\theta_i)$.

$\sum^{n}_{i} (\theta_i) . \leq 2\pi$.

(i.e. Closed <u>Ring</u>).

Seven Important Notes And Results:

The Philosophy:

Quantity $\xrightarrow{}$ **Quality (Quantitative Jump).**
$\xleftarrow{}$

"The waves of water we observe in the oceans or those made by the desert sand is simply Quasntity attempting to be different quality (The Waves)" **The Red Monk.**

The Physics :

"Quantum effects are observed not <u>Only</u> Due to their <u>Microscopic</u> sizes! But <u>Also</u> Due to the <u>Prolificacy</u> Of the <u>Copied</u> Observables at the macroscopic level "

"According to <u>My</u> (Principle of Prolificacy) Stated earlier the atom is confined to the potential well between (–x And +x) Not just because of its microscopic size but also because its crowded (<u>Squeezed</u>)By other <u>Copies</u>."

The Red Monk

1-

I used the terms <u>Gluing</u> and <u>Cutting</u> Instead of <u>Attraction</u> And <u>Repulsion</u> To avoid any confusion with other concepts used in other disciplines of science e.g. electricity.

Since the terms I used will be describing the dynamics of the process of gluing and cutting !Not the <u>Intrinsic Properties</u> of the object itself ie different to that of electrons always negatively charged while protons positive etc.

2-

Every stable object in the universe must have been put together or (<u>Glued</u>) By certain energy at certain frequency (ω_{Gluing})Macroscopically! Or microscopically (<u>The Compton Wave length</u>)! And can be separated or (<u>Cut</u>)With another energy at different frequency ($\omega_{Cutting}$). The difference between these two frequencies we call (<u>Stiffness Factor -Δ</u>).

3-

By reducing the energies in to frequencies we simplify the relationship between the microscopic forces and the macroscopic **Van Der Velt** Force (V) Where:

$$V = 1/r^6 - 1/R^{12} \approx 1/n^2$$

4-

These terms apply to all scales **Macroscopic !Microscopic ! Molecular !Atomic !Subatomis !Or Nuclear Scales!** e.g.

$$\omega - \omega \quad \Delta\omega$$

If : Gluing   Cutting =

Then for proton (The most stable object in the universe);

$\Delta\omega = 0$! For Neutron $\Delta\omega = \alpha$ For Pion $\Delta\omega = \beta$ etc And Van Der Veldt Molecular Forces $= [\Sigma^n \Delta\alpha, \Delta\beta, etc]^{1/n}$.

5- The mechanical energy required to join the parts of any Object in the universe is equal to the energy required to Separate its parts !But the total sum of the energy required to join or separate its microscopic parts is not the same as that required to separate it or to join it macroscopically !

The difference is shown by the following set of equations:

If we have an **Ensemble** Of **Mechanical Frequencies** (f) Rotating around with arbitrary radius (r) Then : **The Probability Of Classical Orbits** =

$$\frac{2\pi r f_1 + 2\pi r f_2 + \text{-----} 2\pi r f_n}{2\pi r f_1 \times 2\pi r f_2 \times \text{--------} 2\pi r f_n} = \Psi_{(f,t)}$$

Or: **The Probability Of Classical Orbits** =

$$\frac{(2\pi r)^{1-n} \sum^n f_n}{f_1 f_2 \text{----------------} f_n} = \Psi_{(f,t)}$$

6--<u>Frequency As Signature For Energy. Frequency in itself is not energy! But</u> it can be an accurate signature for any energy: Microscopic or macroscopic? Celestialphysical or even biological??As in the following examples:

Energy ∕ ℏ(Plank Constant) = 1 ∕ Time(The Period) = ω

7-Calculating My Proposed Universal Constant (H).

"On the expense of shedding some information we <u>Simplify</u> By reducing the energy terms in to those of frequencies! Then we <u>Reconstruct</u> These terms by the expressions for (ψ) Described above:

Thereafter we can actually obtain more informative general view. <u>And That Changes In The Frequencies Can Become A Tool To Determine Changes In The State Function</u> (ψ)."

"Although I called my proposed (H)A constant ?Its organically linked to the classical probability of (f)!!i.e. It's only a constant within <u>Range Of Frequencies</u>(f_n) Determined by $\psi_{(f,t)}$!I am denoing it here as H^*(H-Star Or the conjugate of (H)Where ($HH^*=1$)."

Therefore:

$$H^* = H(\psi_n) \longrightarrow H_n$$

Proceeding From Two Premises:

a-

The **De Broglie Hypothesis** The Wave-Particle Duality. I.e. To avoid complex but quite possible **Relativistic** calculations for this purpose we assume that: If a wave can bend around an object so does the particle.

b- The **Eherfests Theorem**:

$$m \frac{d<x>_\psi}{dt} = <P>_\psi$$

And:

$$\frac{d<P>_\psi}{dt} = m<dv/dt>_\psi = <F>_\psi$$

F-Force.

Imagine there is light source of (Photons) Or a sound source (Phonons) **Emitting** With frequency (ω) ?Now if these sources are caught in orbit of radius (R) Describing **Rotational** frequency (f)? Then we can have two simultaneous expressions for the energy:

A-   Energy = h X ω

B-   Energy = H* X f

Where both (h)And (H)Are as defined above.

Equating both expressions by decreasing the (ω)As far as possible and increasing(f)As much as possible until both (ω) And (f)Are in the same <u>Region Of Magnitudes</u> (Not necessarily identical)Then we have :

$$h/H^* = f/\omega$$

<u>Universal Resonances(PIr).</u>   Part-One.

Since both both (f) And (ω)Are measurable and (h) Is already known we can determine the value of (H).

In the limit when:

f = ω (Universal Resonance)

We obtain: h = H*.

Note:

"In <u>Part-Three Of Universal Resonancve</u> there will be More subtle and less vague ebvaluations of (H*.)"

$$H^* = H \cos\theta = \tfrac{1}{2} h.(\Omega_0/f)\{1- (\phi)^2\}$$

The Coupling Equation Of Classical And Quantum Mechanics.

"If we pick any atom (Or its Isotopes) From the chemical table emitting the <u>Lowest</u> Possible spectra of frequency (ω)? Then if we place this atom in mechanical (Macroscopic) Orbit (e.g. the Cyclotron) Rotating with radius (R) At frequency (f) !Then in principle we can increase this (f)By reducing the radius until $f = \omega$. Since in theory at least as :
$R \text{---------} \to 0.\ f \text{------} \to \infty$ *! Where (f =velocity /R)."*

Quantum.

$Energy^f = ^f h_{(Plank\ Cons}{}^f{}_{tant)} \times \omega$. For <u>Microscopic</u> Objects.

$\Psi_{(\omega,t)} = Re\ [\ e^{i(\omega.t - k.x)}\ ] = Cos\ (\omega.t - k.x)$

Classical.

Energy = H* X f   For <u>Macroscopic</u> Objects.

$\Psi_{(f,t)} = \underline{(2\pi R)^{1-n} \sum^n f_n}$ = <u>The Probability Of Classical Orbits</u>
$\qquad\qquad 1\ \ 2 \text{----------------} n$

Both of the above two equations can be **Quantized** by (u^). As follows:

Diagonal Unitary Representation Of Perturbation Theory: At d ≠ 0 But d² = 0.

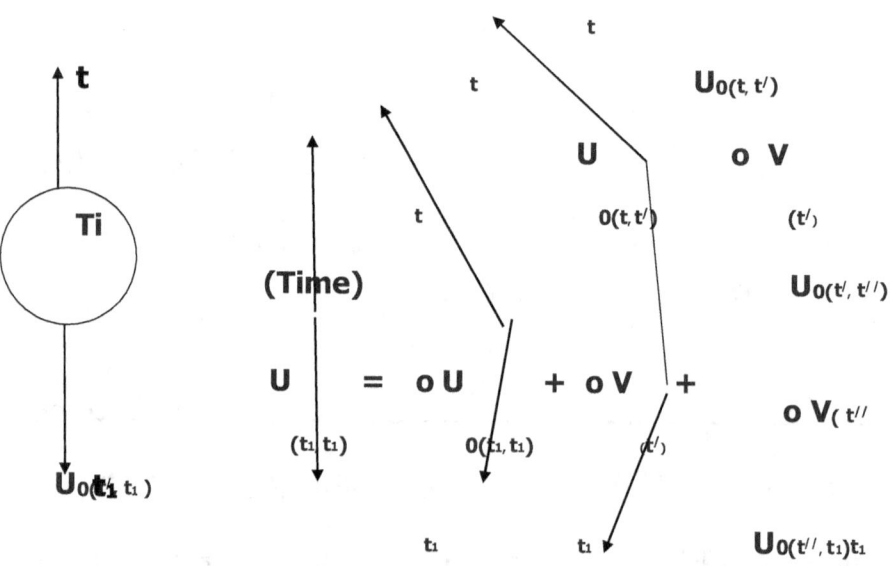

Where ::

$$ih\, \partial \psi/\partial t = H \cdot \psi = (H_0 + V)\psi$$

H = The Hamiltonian . $\Psi(t) = U(t)\psi_0$

$ih\, \partial U/\partial t\, \psi_0 = H \cdot \psi_0$
$ih\, \partial U/\partial t = H = (H)_0 + V(t')$
Lowest order : $\quad ih\, \partial U_0/\partial t = (H)_0$

$$U_0 = e^{-i/h(Ha)0 \cdot t}$$

$$U = e$$

But we said above ; *In the limit when :* $f = \varphi :$ $\quad h = H^*$.

Therefore ; $\underline{U_0 = e^{-i/H(Ha)0 \cdot t}}$

$$[- h^2/2\mu\, X\, \partial 2/\partial r^2 + V_{(r)}]\psi_{(\varphi,t)} = (H)^2 (U_0 \cdot \psi_{(f,t)})$$

$$= H^2\, e^{-i/H^*(H)0 \cdot t} \cdot \psi_{(f,t)} \quad \text{---------} Equation\ (A).$$

Where:

H* = My proposed <u>Universal Constant</u>.

μ = e/2m$_e$ C X M.

C = Speed of light.

M = Angular Momentum.

e = Zeeman electron.

V = The Potential

Ψ$_{(f,t)}$ As before := $(2\pi R)^{1-n} \sum^n f_n$ Equation –B.
$$\phantom{Ψ_{(f,t)} As before := (2\pi R)^{1-n}} 1 \;\; 2 \text{-----------} n$$

Now let us make the following anzast:

$$\{-h^2/2\mu \; X\partial_2/\partial r^2 + V_{(r)}\}\Psi_{(\omega,t)} = (H)f^2 \{1/\sqrt{2}\pi \int - e^{-t^2/2}\}dt.$$

Equation –C.

You will recognise the expression on the right as the normal probability integral?

Back Tracking.

The above <u>Result</u> Can be proven with the methods of <u>Back Tracking</u> By the following steps: (Keeping in mind these steps are just guidelines or hints)! The formal proof itself is rather <u>Long And Complex</u> Therefore delegated to my next book!:

a-

Frequencies offer the <u>Lowest Common Denominator</u> between classical and quantum mechanics.

b-

We need frequencies because they are both <u>Cyclical</u> and <u>Physically Measurable</u>.

c-

We chose above the Zeeman expression for the energy on the left because <u>Magnetic Waves</u> Offer the <u>Lowest Possible</u> range of frequencies (ω_n) Nearest to the region of the <u>Highest Obtainable</u> Mechanical frequencies (f_n) That can be reached. At least In order of magnitude.

d-Since it's not possible to determine <u>Simultaneously</u> the microscopic (Quantum) And macroscopic (Classical) frequencies (ω) and (f) : We need an <u>Ensemble</u> of frequencies <u>At Least In One</u> of the two sets say (f_n)!

Now back to the **The Probability Of Classical Orbits** :

$$\Psi_{(f,t)} = \frac{(2\pi R)^{1-n} \sum^n f_n}{f_1 f_2 \text{---------} f_n}$$

e-By further tweaking of the terms in equations Then substituting in to(B).

f-Also we make use of <u>Fourier Transformations</u> :

$F(t) = 1/\sqrt{2\pi} \int e^{itx} f(x) \, dx.$

$F(x) = 1/\sqrt{2\pi} \int e^{ixt} f(t) \, dt.$

Notice how the term $(1/\sqrt{2}\pi)$ Above corresponds to the L.H.S. of equation –C ?

g- <u>Smoothing</u> The variables since $[df^{(1-n)} = dt^{(n-1)}]$. At this juncture we can <u>Either</u> follow the route of <u>Integrating By Parts</u> ? <u>Or</u> by the following approximation:

h-

Keeping in mind:

$d\, e^x/dx = e^x$ **And**

$\int e^{zx}\, dx = 1/z \cdot e^x$

$\int f(e^{zx})\, dx = 1/z \cdot \int f(zx)/e^{zx}\, dx = 1/z$.

i-

If **(E- the Energy)** *And* **ψ(x,t) Is a function of** $e^{-i.E.t/h}$

Then: $-ih\, d\psi/dt = E.\psi$

j-

Now back to Fourier transformations ; By substituting and tweaking equations (A,B,C,D,E,etc) Should yield the required results.

$\psi(x,t) = \int e^{ikx} \cdot e^{i.\varphi.t}\,]\, \psi(k, \varphi)\, d\varphi\, dK \longrightarrow$

$h^2 K^2 + V(i\, \tilde{\underline{V}}_k)\, \psi(k, \varphi) = ih\, \varphi\, \psi(k, \varphi).$ *Equation -D.*

Escape Frequency:

"We all familiar with the value of velocity required to escape earth gravity. I am proposing here to replace this velocity (V_E) by <u>Escape Frequency</u> (ω_E)."

Since by multiplying through with any arbitrary radius(r)We get:

$$r.\omega_E = r.\omega_{Cutting} - r.\omega_{Gluing} = V_E.$$

Then by removing the (r) we are left only with frequencies again:

$$\omega_E = \omega_{Cutting} - \omega_{Gluing}$$

Defintions:

Hence we now can define the escape frequency as:

"It's the frequency required for any object to escape its own surroundings"

"By degrading (Reducing)The <u>Energy</u> terms in to that of <u>Frequencies</u> we establish a <u>Tunnel</u> Common to both <u>Quantum</u> and <u>Classical</u> Mechanics .But first we need to explore the following hints or tools:"

New Dimensions.

1-The Rock Bottom Common Denominator for both classical and quantum mechanics naturally is (x,t)!

2-Notice the L.H.S. of equation (C) Above contains Spatial term ($\partial_2/\partial r^2$) Which does not correspond to the terms of Normal Probability Integral on R.H.S. of the equation.
Therefore I am introducing here New Dimensions Both are now Scalars And both constituting the very Next Possible Lowest Common Denominators For both Quantum and Classical Mechanics ;The inverses of (x,t) Are :(1/x ,1/t).

The first (1/x) Defined as sigma ($\eth$)= Space Densit(Another Scalar).

While the second (1/t) Recognisable as the Frequency (f)._n_

3-All frequencies (f_n).are Scalars i.e. can be added or subtracted numerically .Similarly space densities ($\eth_n$) Are also scalars.

$$\psi(\eth, f) = N.e^{-ihf} \quad \text{-----Equation -E.}$$

2-

The analogue to __Velocity__ In these new dimensions which I am calling :

$$\text{The Flow} = d\mathring{}/d f_n$$

And The the analogue to __Acceleration__ In these new dimensions:

$$\text{The Blow}_n = d_2\mathring{}/d f^2$$

3- *The Graph:*

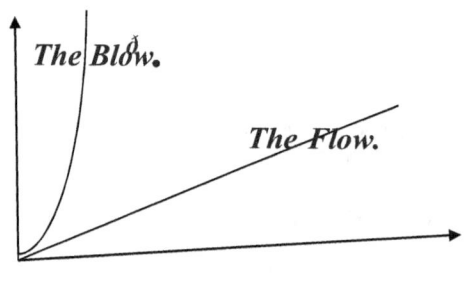

Dimensionally:

$$d\mathring{}/d\, f_n = t^2/x^2 \, (dx/dt).$$

The Flow $= T^2/L^2 \cdot L/T = T/L$.

$= 1/$ **Velocity**.

The Blow $= 1/$**Acceleration**.

Force $\times L = ML\, T^{-2} \cdot L = M L^2 T^{-2} = $ **Energy**.

Mass $=$ (**The Flow**)$^2 \times$ **Energy**.

Mass $=$ (**The Blow**) $\times$ **Force**.

4- **Substituting All The Above Results In Equation (A-E)etc.**

$$\{-h^2/2\mu \times (d_2\mathring{}/d f_n{}^2) + V_{(r)}\}\psi(\mathring{},f)$$

$$= (H)f^2 \{1/\sqrt{2}\pi \int - e^{-t2/2}\}dt.$$

After Applying The Above Terms:

$$\{-h^2/2\mu \times (The\ Blow) + V_{(r)}\}\psi(\mathring{},f)$$

$$= (H^*)f^2 \{1/\sqrt{2}\pi \int - e^{-t2/2}\}dt.$$

Result.

"This is extremely significant result for calculating the Blow at Universal Resonances."

Concrete Results.

Now collecting most of the earlier stated equations:

a- From the prolificacy principle (One, Two, And Three).

n = The minimum number of <u>Microscopic Objects</u> Inside any macroscopic unit.

N = The minimum number of <u>Macroscopic Copies</u> required to observe any quantum effects at distant (D).

h = Plank's Constant.

H = My proposed <u>Universal Constant</u> ????

Then:

$$h \times N = H \times n.$$

b- $$h \times \omega = H \times f$$

c-

And from the second equation of motion discussed earlier:

$$[\Delta\omega / \Delta\omega_{G,C}]^2 \approx -(db^n/db) \approx a$$

Where (a&b) Are any adjacent sides of the <u>Isamic Diagrams</u>. Gives values for (ω & $\omega_{G,C}$) In terms of the spectral frequencies at (a & b).

d-_____ **Critical Hints:**

Diagonal Representation Of Perturbation Theory:

As before : $\underline{(2\pi r)^{1-n} \sum^n f_n} = \psi_{(f,t)}$ **At d ≠ 0 But d² = 0.**
$\quad\quad\quad\quad\quad\quad\; 1\;\; 2\text{------------------}\; n$

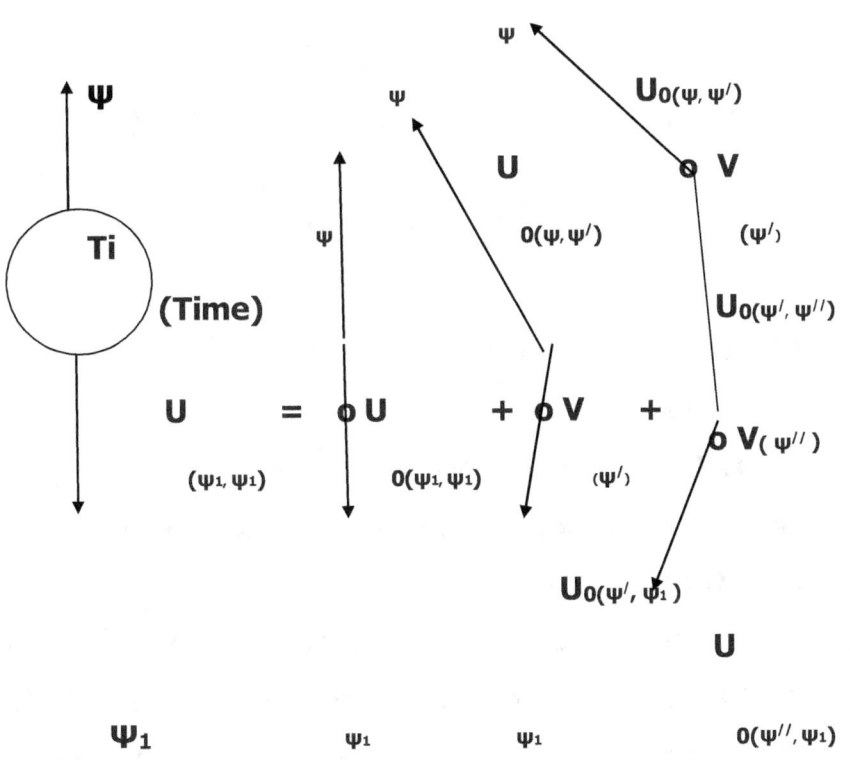

e-

Now equating equations one and two above :

$h \times N = H \times n.$

$h \times \omega = H \times f.$

Giving :

$$N/\omega = n/f = h/H.$$

But : $n \times N = \Sigma n$ Substituting back for (N):

$$\Sigma n / n^2 = \omega / f = H/h.$$

And from equation (4) This is :

$\Psi = K [\omega / f]$. But :

$N = K \cdot Z (A).$ From above where :

N=The <u>Minimum</u> Number of copies required to detect quantum effects.

A=The size of the smallest atom in that chosen unit .

Z=Size of this object measured in multiples of (A).

(This is <u>Not The Same</u> as the number of atoms inside this object.) K=Some constant.

Often it's very difficult if not impossible to measure the energy of objects under indirect or even direct investigations!

However it's possible to measure their frequencies!

Frequency in itself is not energy but by converting energies in to frequencies then comparing an ensemble of frequencies each treated as a sign or signal of the corresponding energy ! We can establish the energy by eliminating the constants in the equations.

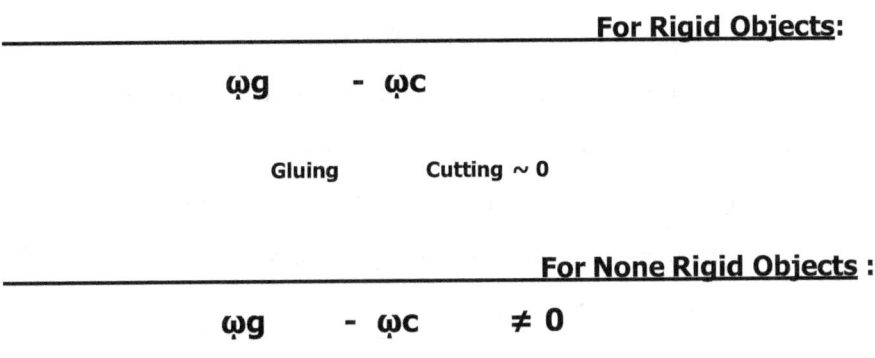

For Rigid Objects:

$$\dot{\omega}g \quad - \quad \dot{\omega}c$$

Gluing   Cutting ~ 0

For None Rigid Objects :

$$\dot{\omega}g \quad - \quad \dot{\omega}c \quad \neq 0$$

Gluing   Cutting

The practical significant of the above two equations will be revealed in my next book. e.g. By introducing the <u>Stiffness Factor</u> above and the differences between the frequency of (<u>Impact</u>)) And that of (<u>Recoil</u>) We can measure the size of photon from its momentum ! And other known properties.

In The Laboratory.

"Where to look in practice for these <u>Universal Resonances</u>?"

1-

Similar to <u>Electrical Batteries</u> There are <u>Mechanical Batteries</u> To restore excessive mechanical energy!

These are cylinders suspended inside casings of magnetic fields rotating frictionless at incredibly high speeds!!

They are kept underground not for any sinister reasons but only to avoid disastrous accidents if any of them break away!

Spinning at such extraordinary speed can cause damages equivalent to the explosions of thousands of TNT ! ? !

If their angular velocities start touching the lowest range of spectral frequency then the phenomenon of <u>Universal Resonance</u> Should be observed!

2-Another place to look for are the <u>Cyclotrons</u> :The particles inside these are accelerated to incredibly high speeds where their linear velocities can reach the <u>Speed Of Light</u> But unfortunately their rotational angular velocities are not so high yet ?

In Search Of Scientific Definition Inside Non Scientific Subjects.

*Before entering the next section discussing the **Constant Of Destructivity** with **Scientific Scales** We need to seek **Scientific Definition** for **Destructivity**,*

As we said in **(Red Monk Law—See Page)?** *We cannot define **Morality** but we certainly can define **Immorality** and we gave the example of murder its immorality or criminality which can never be disputed by any two(A)&(B)1 But let us assume that A Wrong B Then Why? and how ? is debatable ! If A kills B Then it can be still debatable (Was it in self defense ? Was it provoked ??etc.) However if A Kills **Hundreds** then the dispute turns to certainty The **Certainty Of Wrong Doing** and this is what makes any definition **Irreducible Scientifically** i.e. valid for all season*

*But notice we said if A kills hundreds meaning the **Figure Of Hundred** is what brought the certainty about ! **Hence Figures And Numbers Plays Central Role In Determining How Scientific Our Definitions Are** ? As we shall see later :*

Inescapable Responsibilities.

"Here The Quantity (Numbers) Determine The Responsibility Beyond any reasonable doubt."

"Once in public debate about the Holocaust on afro patriot I believe he was from Nigeria got up and angrily shouted ((What about us ? Nearly hundred millions killed in sub-Sahara Africa by British colonialism according to respectable reliable international organizations and not a sound about it ! Are we too not humans ? Or do we not count like the Jews???"

As you see Facts And Figures Not propaganda propelled emotions must play central role in our definitions ! No one can dispute the Immorality Criminality Or Sheer Destructivity of deliberately and Systematically making hundred millions perish.I.e. no excuse will be enough to evade responsibility.

Acts of genocides or Sheer destructivity can be Temporarily white washed by sophisticated lies or expensive propaganda but History Operate Differently Once the figures had become too massive to be covered up :**(See Page---The British Art And Science Of WhiteWashing History) ?**

Another example ((keeping in mind)) its only one

example from thousand plus similar atrocities committed by the British)) is the :Jallianwala Bagh Massacre -1919.

Jallianwala Bagh is a garden temple where the Indians gather once a year to pray and to chant religious songs.

The English general (Dyer) Came in with troops lined his soldiers

up and ordered them to fire to kill the peaceful worshipers as result thousands died or injured in this totally <u>Uncalled For Massacre By The English.</u>
The Indians much later on forced court hearing by which the English general claimed that the worshippers tried to rush his troops yet his own bodyguard and sergeant testified <u>Under Oath</u> that
none of them were armed and none of them tried to rush us!

in fact they were totally indulging in peaceful prayer not even aware of our presence until we fired .Now the reader may be in position to accept <u>Figures</u> as concrete guide for our definitions :

For example if we add up the:

<u>Total Destructivity Of The British = Killed Thrty Five Millions In The Indian Subcontinent (The Partition)+Ehty Seven Millions In Africa +Eleven Millions In The Mifddle Est (Including The Iraq War)+One Million In Ireland (Perioshed In The Infamous Potatoe Poisoning) = 132 Millions (At The Very Least Victims Of English Colonialism)</u>

On the other hand if we add up the Nazi crimes of :

The Total Destructivity of The Nazis =6millions Jews +27 Of Our Slavic Brothers +2millions Others =35 Millions

As you see the crimes of destructivity by the Nazis will fade away when compared to the British crimes against mankind ! That is why the English will do everything to kill any Scientific Treatment to these crimes knowing it will then be extremely difficult for them to wriggle out of it !

To extricate themselves by sophisticated lies! show trials! and expensive propaganda.

That is why the English will never allow such Concrete Scientific Yard Stick To develops that will Pin Them Down (Identify Them Indisputably) as the Sole Global Criminal Enemies Of Mankind ! All Of Mankind..(See Page---The Enemies Of Mankind)

3-On The Self Destructivity Of Man And The Constant Of Destructivity (Later)

Or To Borrow The Concept Of (Impulse) From Physics:

Where

D = Force of destruction

T - Time the force is applied.

$$D_{(Colonialist)} \ll D_{(Fascist)}$$

$$T_{(Colonialist)} \gg T_{(Fascist)}$$

$$\int_{t1}^{T} D(Fascists) = \int_{t2}^{T} D(Colonialists)$$

****

Introducing The Heterogeneous Units Of Measurements.

"Playing acrobatics with logic! Can yield new insights to unitary dimensions."

" One Apple +One Orange = Two Fruits."
Two Fruits. +One Sandwich = Lunch."

"At first sight this book may seem to be cavalier (Easy) About <u>Physical Units</u> !Its not !!It just that any equation can be multiplied through by any constant which can transform the physical units ."

All dimensions can be transformed from the conventional to <u>Purpose Built Dimensins</u> :

To illustrate this point we take as <u>Example Only</u> :The equation of the constant of destructivity which keep appearing at many parts of this book in various context and formats :

$$D \times T = T \times D = \text{Constant}$$

The constant in the equation is kept constant by <u>Exchanging</u> the following three :

A-

Their position in the equation i.e. :

$$D \times T = T \times D$$

B-

By exchanging their absolute numerical values i.e. :

$$3 \times 2 = 2 \times 3 = 6.$$

C –

By exchanging the units of each i.e.:

If the units of D = Number of people destroyed in millions.

And

T = Time in years

Then :

Destruction (D) . Years (Y) = Years . Destruction (D).

Or:

DY = YD = Constant.

And Since Its Constant :

This Can Be Scaled Down To Units Of YD Or DY.

And the above example becomes:

3 D X 2Y = 2Y X 3 D = 6 DY = 6YD.

Applications.

Imagine a black box (Of unknown) But <u>Quantified</u> particles (D)? And we need to do work or obtain work from them (e.g. by moving) in finite time (T)! Then:

$$[D]_{d1}^{d2} \times [T]_{t1}^{t2} = [T]_{d1}^{d2} \times [D]_{t1}^{t2}$$

Or Simply :   **D X T = T X D**

Where: d1 = 0   T1,t2,d2 ≠ 0

If two are known the rest of the four can be determined If one is known and the other can be estimated i.e.

<u>**Quantified**</u> *the rest of the four can be determined!!*

See later for the difference between <u>Quantifiable</u> and <u>Quantifiable?</u> Now We Quote Three Very Fundemental Applications A?B ?And C:

If two are known the rest of the four can be determined

If one is known and the other can be estimated i.e.

Quantified *the rest of the four can be determined!!*

See later for the difference between Quantifiable and Quantifiable?

<u>**Now We Quote Three Very Fundemental Applications A?**</u>
<u>**Now We Quote Three Very Fundemental Applications A?**</u>
<u>**B ?And C :**</u>

<u>**Now We Quote Three Very Fundemental Applications A?**</u>
<u>**B ?And C :**</u>

A-Speed Of Light (Part-Two)   Dimensionless Analysis.

"According to G.R time and space can intermingle (Exchange Units Or Dimensions) Under certain conditions (e.g. Black Holes).

By switching notations the following two black boxes effectively become our black hole.
From one box we measure the distance moved only!
From the other we measure the time taken only!!
Per say per same specific amount of effort"

If we move inside the above box **Cluster** of particles (D) In time (T)**Head on** against something e.g. another cluster then by denoting each cluster with d1,d2,d3,etc

Then if we have another identical black box of **Identical But Still Unknown** Particles?

*By subjecting it to certain identical effort e.g. work **Already Measured From The Previous Box** we can dispense with all the (d)s by denoting each cluster this time by (t1, t2. t3 , etc) Instead of (d1,d2,d3,etc) i.e. by substituting it with time(t) **Measured Outside The Second Box**.*

*Thus by moving one **Cluster** Against the other we can establish the number of particles moved:*

*But the most interesting result is when the process is inverted **Again** By knowing the (d)s this time e.g. number of photons we can actually tell how long it will take them to move **Without Measuring The Time Itself** ! This can be shown to be another proof of the constancy of speed of light inside this black box But only in **Localized Regions Of Space** .*

B- In Quantum Mechanics;

"We already said above (classically): Only one of the four need to be determined to estimate the rest."

$$\bar{a} = \frac{\int D^*(t)\, a_n\, D(t)\, dt}{\int T^*(d)\, T(d)\, \partial(d)} = a_n$$

The choice of the letter (d) Was unfortunate but clearly d(d) mean The diff of (d).

Where

a_n =The average.

D, D^* =The state and its conjugate.

ΨC-

Taking this idea further let us now imagine our constant this time is a cake (Denoted by the letter -C) Made from the following ingredients:

2 Eggs +3 Apples + Oranges. etc.=Cake "C".

At <u>one stage of the logic</u> we can say that one egg makes only half cake and one apple a third of the cake! etc. Therefore this can be represented as follows:
A cake is made of half cake plus one third cake +etc.
OR:

C= ½ C+1/3C +1/4 C= 13/12 C

Generally speaking

For Tree

T =x Carbons+ y Nitrates+etc. OR:

$$T = \frac{x+y}{XY}$$

And the Carbon $=1/x$ while
the nitrates $=1/y$ ----- etc.

Applying group theory we can find the new unit for any element by a permutation of intersections representing the <u>Identity</u>? Just the way a single point can be identified by the intersection of <u>ANY</u> two lines drawn through it ? ! ?Most units of measurement are of predetermined parameters?

e.g. that of force is simply units of

$$\text{Mass} \times \text{units of acceleration.}$$

$$1/F = 1/[\ F\ \text{Mass.} + F\ \text{Accelerations.} \quad]\ \text{etc ??}$$

$F = ma$

$$1/m + 1/a = \frac{m+a}{ma} = \frac{m+a}{F} = 1.$$

At unity $F = m+a$. or :

$F = ma$ ----------------------- $\rightarrow F(m+a)$

Similarly for other units of measurements.

As to what purpose this equivalence employed is not of any concern! As long as it is homogeneous? What I am proposing here is **Backtracking** on this! i.e. first we determine the **objective of measurement** then organically link this to the units! For example? Let us introduce an [x arbitrary units] of whatever it takes to produce an apple tree? And this is equal to "Y" units of calories required plus "Z" -units of sunshine or whatever **catalyst** is required? Therefore;

One "apple- like unit of measurement" is [$1/x = 1/y + z + etc.$]

Similarly in social domains we can say:

"Revolution -like units" = $1/$ [K-number of population +Y-number of rebels +S-units of sunshine etc?]

(Believe it or not the Russian October revolution happened in November after **Exceptionally Long Hot Russian Summer!**"

Needless to say that this system of measurements will be organically linked to the objective? And **varies from one objective to another! i.e. .**

BOTTOM LINE-S.

"The truth like positron dies the moment we discover it" **The Red Monk.**

"When all the curtains of lies are drawn! The truth remain dancing solo! And singing soprano!" **The Red Monk.**

"Like deaf and dumb(Village Idiot) in a Brazilian carnival the truth inside Britain is dancing solo " **The Red Monk.**

"Even a lie is a piece of information! Will tell us something about the liar" **The Red Monk.**

"The easiest thing in the world for the ignorant-s of the world to dismiss anything incomprehensible as paranoia ." **The Red Monk.**

The Quantified ? Quantifiable And the Quantifiably .

"Here we express these three with <u>Socio –Political Parameters</u> ! For <u>Physical Parametrics</u> See chapter seventeen and eighteen."

The Quantified.

If we need to make list of countries to compare their deviation from what is considered to be (Normal) ? :

We can choose one singular anomaly of (Any Available Reliable Statistics)e.g. The number of homosexuals inside the nation then dividing this number by the population of that nation !Simple. And this is <u>Quantified</u> .

(See Page----Degrees Of Anomalies)?

The Quantifiable.

When colonialists like the British occupied and colonised killing countless millions from <u>Africa</u> to the <u>Middle East</u> to the <u>Indian Subcontinent</u> and back to the <u>Middle East</u> !They could not have achieved all of that by charging their own population with love or tolerance but with hate and only hate which I shall call hereafter <u>English Venom</u>.

This hate or venom cannot be measured directly but only by its consequences? Simple but concrete scale of <u>Direct Proportionality</u> i.e. it will be a <u>Function</u> of the number of people killed by the British during any given period(<u>Divided</u>) By the population of Britain during that period of time.

I called this quantity the B.U.V.O (<u>The British Unit Of Venom</u>).

And if we repeat the same procedure for the rest of Europe calling this result the E.U.V.O . for (<u>European Unit Of Venom</u>).

I found out that the : BUVO >>EUVO   (Always much larger .)

The Constant Of Constant-s..

What is most striking about these constants the BUVO And EUVO That their ratio <u>Remain Constant</u> Even during peace times ! ? !

This can be proven as follows:

If we take the number of <u>Overseas Students</u> or people in general of <u>Foreign Origin</u> who had entered Britain since the end of the second world war and who had been deliberately <u>Disabled</u> or had been stealthily sentenced to premature <u>Deaths</u>(As described by this book) Then dividing these numbers by the average population of Britain since the second world war denoting this as (<u>Peace Time- BUVO</u>). Then repeating the procedure with other (Normal)European nations e.g. France or Italy denoting the result as :

(<u>Peacetime- EUVO</u>). To our shock and horror we discovered that : The ratio of : BUVO(<u>Peace Time</u>)÷ (Divided)by the EUVO (<u>Peace Time</u>)= BUVO(<u>War Time</u>)÷ (Divided)by the EUVO (<u>WarTime</u>) = Constant! ? ! And these are the <u>Quantifiable</u>;

(See Page-----------------------------The BUVO And The EUVO) ?

(See Page-S -------------------Silent Genocides Parts 1,2,3,Ec) ?(See Page------Death Squads Parts 1,2,3 ,Etc) ?

The Quantifiably .

We can not measure *Morality* directly but we can measure *Immorality* .For example *We All Agree* That murder is immoral (To say theleast)!Therefor the measured degree of *Morality* is the inverse function of the degree of *Immorality* Already measured from calculating the *Number Of Murders Per Say* .

(See Page----The Red Monk Law) ?

(Seepage-----------Let the Statistics Speak For Itself)? Also :(See Six Calls To The United Nations On Pages---) ?**his Was Only The Summary !For The Rest See VolumeThree(the Five Arithmetic-s By Visiting** :www.scribd.com/isamtahersaleh

In This Book We Only Allow Numbers Figures And Equations To Speak Out.

"Here we list <u>Most</u> of the equations <u>Derived</u> by this book."

1-   The Constant Of Destructivity.

Let:

d = the absolute or total destruction.

T = Time it had taken.

D = Destructivity or the rate of destructivity D = d/T.

Then for any two entities (A & B) :

We have:

$$D_A \times T_B = D_B \times T_A$$

Dimensionally :

$$d_A / T_A^2 = d_B / T_B^2$$

827

Which Is Consistent With The Graphs Of Decay And Destruction:

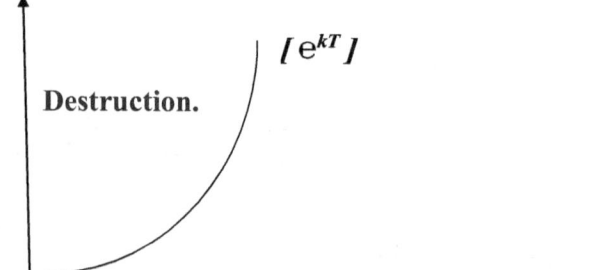

$[e^{kT}]$

Destruction.

Time.

Now if we **Plug** In all the available figures (Given inside this book or by any other **Independent** Source) For:

A=Germany.   And   B=Britain.

We arrive at one and only one conclusion that is :

The Total Destruction (d)By Britain Is At Least Ten Times Bigger Than That Of Germany .

_____Curiously.

If we now substitute the units of distance for (d) Instead of (Destruction) And add (d)To both sides of the equation then multiply both sides by (d) Sigma = Mass per unit length :We get :

$$d \cdot d/T^2_A + (d) \in d = d/T^2_B + d \cdot \sigma$$

Which is **Analogous** to the all too familiar equation of :

$$\text{Force} = \text{Acceleration} \times \text{Mass}.$$

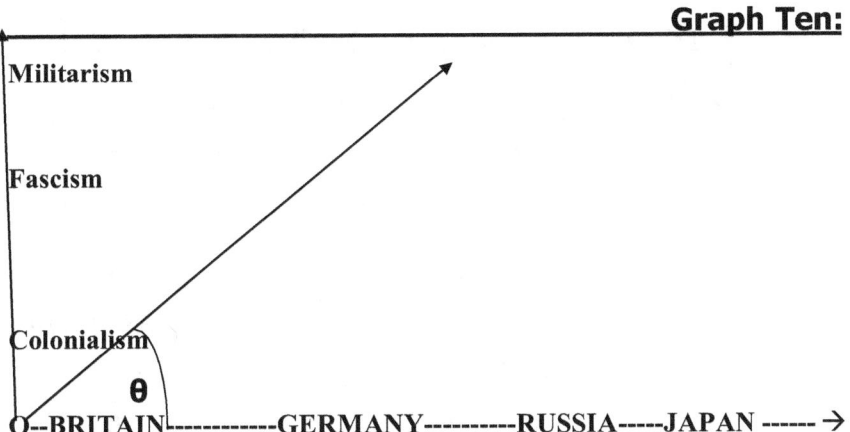

Graph Ten:

Where (θ) Is the constant of destructivity applied in the equations derived earlier as follows :

1-   The Constant Of Destructivity.

Let:

d = the absolute or total destruction. T=Time it had taken.

D = Destructivity or the rate of destructivity D= d/T.

Then for any two entities (A & B) :We have:

$$D_A \times T_B = D_B \times T_A$$

Dimensionally : $d_A / T_A^2 = d_B / T_B^2$

Further Interpretation Of My Equation On The Constant Of Destructivity And Graph Eleven.

"The most striking part how the proof to my theory present itself <u>Even</u> at the <u>Tail End Of The Argument</u> i.e. <u>Even</u> in <u>Peace Time</u> When the <u>Ratio</u> of overseas students destroyed by the British to the population of Britain is equal to the <u>Ratio</u> of overseas students destroyed by the Germans to the population of Germany and its equal to constant!"

$$[D_1 \times T_2]^Y \text{(Fascists)} = [D_2 \times T_1]^X \text{(Colonialists)}$$

Let :

D_1, D_2 = **The Number Of People Killed <u>Directly</u> By The Fascist And The Colonialists.**

T_1, T_2 = **Time Taken To Kill Thèse People In Each Case By The Fascistes And Colonialiste .**

And Let :

X = **Total World Population During The Period When British Colonialism Began .**

Y = **Total World Population During The Second World War.**

1- First we need to calculate the constant of the equation from <u>Available Concrete</u> Figures for example :

We know for certain now that during the Six years of the second world war i.e. $T_1 = 6$ Years. The Germans (Nazis) Killed <u>At Least</u> Thirty millions (<u>Six Millions Jews Plus Twenty Four Millions Slavs</u>).

Therefore the constant of the equation :

$= D_1 / T_1 = 30/6 = 5.$

2- We also know <u>For Certain</u> That British <u>Direct</u> Colonialism did last (Again) <u>At Least</u> Seventy years i.e. $T_2 = 70$ Years.

3- Now plugging All the numbers above in the equation :

$$D_1 / T_1 = 5 = D_2 / T_2$$

Will give the number of people killed by the British as D_2 = One third of a Billion !

4- So what has gone wrong with this equation because **Surely** the Brits could not have killed **One Third Of Billion** When the population of the entire world was only one billion? ! ?

5- The answer is found in the (**Inflationary Term** (X/Y): Just as one thousand pounds nowadays is equivalent (Purchasing power) to fifty pounds at the beginning of British colonialism similarly: X And Y as defined above: X= One billion world population while :

Y= Four billions world population during the second world war.

Hence introducing the inflationary factor X/Y = 1/4.

6- And the full equation of mine become :

$$[D_1 \times T_2] = [D_2 \times T_1] \, (X/Y).$$

$$[30/6]^4 \text{(Fascists)} = [D_2/200]^1 \text{(Colonialists)}$$

Or:

$$[5]^4 \text{(Fascists)} = [D_2/70]^1 \text{(Colonialists)}$$

Giving :

$D_2 = 5 \times 70 \times 1/4 \approx 80.7$ **Millions People killed by the British worldwide and this most certainly tally with all available recodrs when you add it all up as follows :**

Total people killed by the British :

= **35 millions in the** <u>Indian Subcontinent</u> + **5 millions in the** <u>Middle East</u> + **Eleven millions In** <u>Africa</u> + **Unknown Miscilanous e.g. (** <u>South East Asia</u> +<u>American Red Indians</u> +<u>Australian Aboriginaries</u> **) + etc +etc** ≈ **Eighty millions.**

Please Note The Following :

a-We had taken the safest **First Order Of Approximation** By plugging the **Absolute Minimum** :

This was indicated above by the words (**At Least**).

b-We only counted the number of years (70)As the minimum in direct (**Contact Killings Colonialism**) Inside the **Three Centuries** of British colonialism.

C-_____ **And The Numbers Still Rising**.

What this equation is telling us in **Concrete** terms that the British had murdered the **Equivalent** of one billion people in today's terms (By Today's Scale) !Or just less than **One Tenth** of entire world population in yesterday's terms. Period. However to predict an estimate for the size of atrocities committed by any growing power e.g. the U.S.A. We proceed as before but replacing **Past Period** (X) By the **Future Period** (Z):

$$\left[D_1/T_1\right]^Y = \left[D_2/T_2\right]^X = \left[D_3/T_3\right]^Z$$

With The (Inflationary Period)Now (Z/Y):And So Forth:

(See Page---------- Numbers That Speak Louder Than Words)?

Graph Eleven:

After Superimposing **Graph Seven On Graph Ten**:

With:

$$T_{Colonialists} \gg T_{Militarists} \gg T_{Fascits}.$$

Whereas now : T=Time (The inverse of the periodicity shown in graph seven) And by substituting :

$T_{Fascits.}$ =Six year of Nazi Germany (Second World War)

$T_{Colonialists}$ = Four Centuries of British colonialism.

Etc:

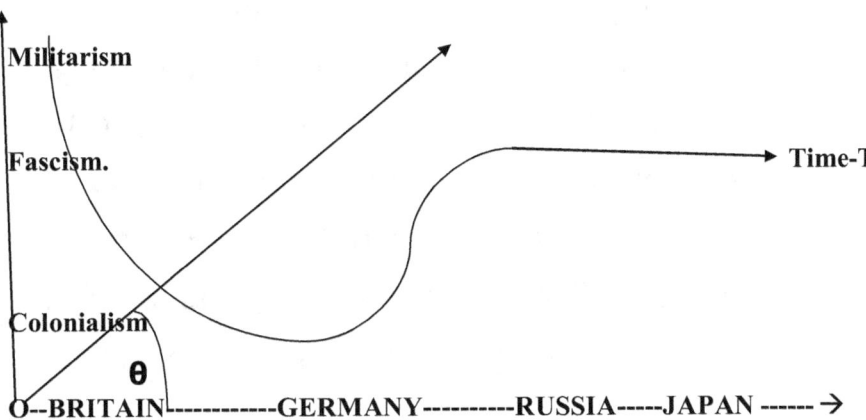

a- We had taken the safest **First Order Of Approximation** By plugging the **Absolute Minimum** :

This was indicated above by the words (**At Least**).

b- We only counted the number of years (70) As the minimum in direct (**Contact Killings Colonialism**) Inside the **Three Centuries** of British colonialism.

C- _____ **And The Numbers Still Rising**.

What this equation is telling us in **Concrete** terms that the British had murdered the Equivalent of one billion people in today's terms (By Today's Scale) !Or just less than **One Tenth** of entire world population in yesterday's terms. Period. However to predict an estimate for the size of atrocities committed by any growing power e.g. the U.S.A. We proceed as before but replacing **Past Period** (X) By the **Future Period** (Z):

$$D_1 / T_1 \quad = \quad D_2 / T_2 \quad\quad D_3 / T_3$$

With The (Inflationary Period)Now (Z/Y):And So Forth: (See Page---------- Numbers That Speak Louder Than Words)?

Intrinsic Propensity For Destructivity The (I.P.D.).

"Our equation now reappears in the form of :

$$(T/d)_A = Constant = (T/d)_B$$

And we define the absolute value of destructivity for any entity (A) As the <u>Inverse</u> Of the <u>Rate</u> of total destruction to determine the <u>Intrinsic Property</u> for destructivity inside each entity (A,B, etc. :

$$D_A = (T/d)_A"$$

If we now plug in the figures (A) For Germany And (B)For Britain in the above equations :

We all know that the Germans were directly responsible for killing <u>At Least</u> thirty million people in the second world war which lasted six years i.e. $D_{Germany} = 6/30 = 1/5$.

And we also know that during British colonialism of <u>Two Centuries (200 Years)</u> the British had killed <u>At Least Thirty Five Millions In The Indian Sub Continent Plus Five Millions In The Middle East And Nearly Ten Millions In Africa</u>.i.e for Britain:
D= 200 Years ./(Divided by)<u>35 + 5 + 10</u> = 50 Millions = 4.

2- Back Tracking.

"Again we apply the scientific methods of backtracking"

We cannot measure the amount of hate required to kill certain amount of people but we certainly can measure the number of people killed by that amount of hate which I call :

Communal Venom.

Hence we define: The Ratios:

BUVO (British Unit of Venom)

= Number of people killed by Britain during any specific period of time **Divided By** the population of Britain during this period.

EUVO (European unit of venom)

= Number of people killed by **Any** European Country during any specific period of time **Divided By** the population of that same European country during this period.

Notice:

Observe the wording (At Least) i.e. The Safe Solid Absolute Minimum In all of these calculations ?

3-   **The Constant.**

"What prompted this study in the first place was the <u>Phenomenon</u> of the inertia of killing (Wogs)By the British extending itself even in to peace times ! **(See Silent Genocides**) ?"

A-

We already calculated the (I.P.D) for both Germany

D$_{Germany}$ = 1/5 And that for Britain D$_{Britan}$ = 4.

Therefore their war time ratio (W) For destructivity was that of Britain to Germany W = 4 Divided by (1/5) = 20

i.e. British destructive is <u>At Least Twenty Times</u> worse than that of Germany .

<u>Curiously</u> This ratio stays the same even inside <u>Conditions Of Peace Times</u> as follows:

B-

<u>Now</u> if we calculate both the BUVO and EUVO as given above in conditions of peace (P)Then take the ratio of the <u>(BUVO) Divided By The (EUVO)We find it</u> Stays the same <u>Even</u> at peace times (See <u>Silent Genocides inside Britain</u>)Where P = 20.?i.e. a constant of the two constants W/ P = 20/20 =1

The I.P.D. The Analogue Of P.E.

"The <u>Intrinsic Propensity for Destructivity</u> (I.P.D.) Is the <u>Potential Destructivity</u> analogous to the <u>Potential Energy</u> (P.E).".

_____ Question:

If we already know the British had killed <u>At Least</u> Twenty times more than the Germans? Then why do we need to redefine (Destructivity)As the inverse of the rate of destructions ?
Why can't we just simply say Britain had killed so many more than Germany ?

_____ Answer :

Yes knowing the concrete figures of the number of people killed by each tell us the absolute truth but does not offer us the <u>Scale or (The Metric)</u> To measure the **Destructivity Of Each** in all circumstances e.g. (War Or Peace)!To apply the **Metric** For any period of history or (Even To Predict The Future) Its analogous to the P.E (<u>Potential Energy</u>) Term in physics! But only in the sense its (<u>Potential Destructivity</u>):

The definition need to be clear and closed (Not open ended) For this we must insert the **Time Dimension** i.e. we need the inverse of the rate of destruction which contains (Time) .

The Metrics Of Relative Destructivity.

We already calculated the **I.P.D** for Germany was (1/5).
And for Britain =(4).
Therefore we said their **Relative Destructivity** is :
(4)/(divided)By (1/5) = 20. I.e.
Britain in **Real Terms** is **Twenty Times** More destructive than Germany! Meaning :
If in **Peace Time** the Germans destroy the life of one **Schwarze** (Wog)Per day ? Britain is destroying twenty .
Similarly it's recorded fact that during the Algerian war of independence the French killed one million Algerians in the last year or two of this war !Hence;
French **I.P.D** =One Year /(Divided By)One Million =1 (Approximately).
And the **Relative Destructivity** of Britain to France will be 4/1 =4.
i.e. if the French again inside **Peace Time** Destroy the life of one **Gypsy Per Day** ?
The British will destroy **Four Times** that much.
And so forth:

_____**Application.**

"In this book we only allow numbers figures and equations to speak for themselves."

From available <u>Figures</u> we can build a list for most of nations similar to those we constructed for the <u>Degrees Of Anomalies</u> :

(See Page------ Degrees Of Anomaly) ?

Which was based on the total number of homosexuals inside the nation divided by the total number of its population!

<u>The Most Striking Discovery Which Comes Out Of Comparing The Two Lists:</u>

<u>They Are Found To Be In Close Correspondence With Each Other!</u>

<u>Proving That Homosexuality Is Just Another Form Of Destructivity</u>.

(See Page ------ Crime And Homosexuality) ?

Angels Of Destructivity.

"To maintain this constant? Destructivity needs its own Permeant agents or angels! Clearly the above equations are telling us who are these angels of Destruction? They are the E.S.P (And Their Jews) ."

(See Page_____-_Killer Race) ?

"By the end of this book it should had become abundantly clear for anyone who read it properly: That given a choice between the fascism of Hitler and Mussolini or the Protracting Indefinite Latent Colonialism of Britain and France ? The first is much more preferable to the latter because at least there was light at the end of the tunnel."

(See Page_____-_____-Super Glue) ?

Memorial Days.

"In addition to the <u>United Nations</u> Declaring January the 27th as memorial day to remember the Jewish holocaust in Europe :

The united nations should in addition (Not instead) allocate another <u>Memorial Day</u> To remember the <u>Countless Millions</u> of (Wogs)Killed and are still murdered (e.g Iraq) By the colonialists in <u>Asia The Middle East And Africa</u> !(So it may never happen again)! Or does the U.N . consider the lives of these (Wogs)Not Of the same value as those of the Jews or other Europeans "

Islands And Penesulas.

"Peninsula =3/4 Island."

For more details on this subject see five arithmetic-s (See Page-------Islands And Peninsulas)?

2-The Graph Of Decay Of Nations.

(The Decay Of Nations Part Four).
Growth.

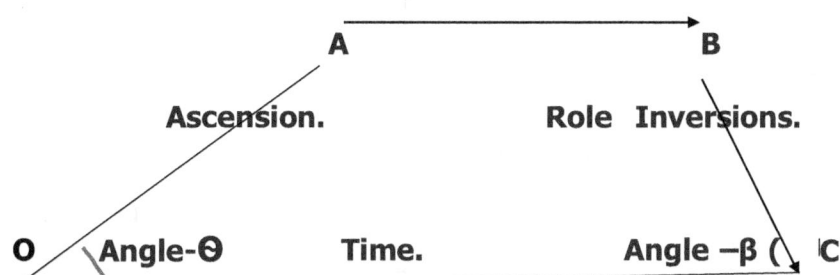

Angle βetta >> >Always Much Larger Than Θ.

Symptoms at (OA) =Fascism Or Famine !
(See Page---Hunger Is The Progenitor Of Their Fascism) ?

Symptoms Of (BC)= <u>*Inversions*</u> *of Rloes ! Of Logic! AndOf Priorities !Even of Sexuality (i.e. Homosexuality).!!* **(See Page---------------------The Pink Pound) ?**

The bigger angle betta the more sever are the inversions in some cases when(Maximum β =π/2) There will be <u>Total</u> inversions e.g. Dr Shipman (See Page--) ?*Or the Grenfell tower fire* (See Page---)?

C.C.G. (Carriers Of The Criminal Genes, What Is Genetics ?

"When a bird is singing ?Flying? Or Picking ?
It did not go to any school teaching it how to do so!
The bird often without any help not <u>Even</u> From its own mother does it driven by <u>Genetics</u> that had been mutated in several generations.
Similarly when an E.S.P. (English Speaking Person) Hates or shoot at black people like clay pigeons (See Page---Clay Pigeon) ?Or committing those <u>Verified</u> Crimes against humanity in <u>Basra ?Abugraib?Etc</u>? (Iraq)!Or the <u>May Lay</u> massacre (Vietnam)!He did not learn it at any school or whatever? He or she (e.g. Ms England of Abu Graib)Are driven by genetics that been

formed during <u>Five Centuries</u> of <u>Ch attel Slavery</u> followed by at least <u>Three Centuries</u> of <u>Direct Colonialism</u>!

It's for these very same reasons you will observe inside nations that did not practice slavery like Russia and Germany <u>Racism</u> is always <u>Short Lived</u> and superficial. For <u>Brevity</u> And since there is no other <u>Motives</u> for these criminal acts committed by the E.S.P(And the "P" Now stands for People) :

Hereafter this book will call these motives as (<u>English Venom</u>)!Or pure English venom."

(See Page---What Is Genetic Mutation)?

(Seepage---The Second American War) ? (See Pages------------Matters Of Science) ?

What Is Genetic Mutations ?

"They tell us how genetics are behind this illness or that disease ?But they dare not tell you how genetics are behind the notorious criminality(<u>Both Institutionalized Or Otherwise</u>)Of the E.S.P.!!"

" <u>Genetic Mutations</u> *Inside human beings do not happen overnight! It takes time !Lots of time !! Lots of <u>Generations !</u>Measured in scales of magnitude its only compatible to the <u>Centuries</u> of British colonialism and before it that period of slavery ! Not the <u>Decades</u> of Nazi Germany or Japanese militarism. That is why racism inside Germany we note is <u>Transient Superficial Open :Therefore Curable.</u>*

*While the racism of the E.S.P. Is <u>Deep Institutionalized Subtle Stubborn Sly Slimy Sneaky Therefore Incurable.</u>"***(See Page----Fundamental Forces)?On Why British Or American Policemen See Blood At The Very Sighting Of Any Black Person ??Also: (See Pages —Clay Pigeons) ?**

Guns do not kill people !Only people kills people! To kill so many people ?You need so much hate. Both the <u>Shock Fascists</u> (The Nazis)And the <u>Protracting Fascist</u> (The Colonialists)In addition to arms and armaments they needed hate ! Lots of hate ! Such amount of hate(<u>Real Or Fabricated</u>) Does not come out of the blue! It needs to be <u>Generated And Sustained</u> Systematically !

A lengthy process of <u>Indoctrination</u>! <u>Arguments</u> ! <u>Even Mythologies</u> (As it's the case with the Israelis)To motivate their own people and industry to kill

Weather this hate can be <u>Quantified</u> Or not is <u>Immaterial</u> Because its consequences are very much <u>Material</u> !

It's the <u>Concrete Number</u> of people killed by the <u>Concrete Number</u> of people who hated them! Period .It really baffles me how such concrete <u>Direct Correlations</u> Between the <u>(Amount)</u>Of hate and its consequences had escaped these scientists so far.

Colonialism From The Objective To The Chronic To The Genetic.

"In the beginning the colonialists colonised by the Objective necessity to obtain Raw Materials !Then after centuries as its the case with the English it became Chronic !! Then the chronic turned in to Genetic Mutations For the criminal ways of life and living !! Again You Do Not Need To Believe Anyone : Just observe how crime and criminals in all the strata of their communities plays central role inside the averagelife of the E.S.P.(And the –P- Now is for person)" **Scientists who deliberately or otherwise missed this crucial point !**

By not providing the required scale for measuring it !

It Seems The Lack Of Honesty Can Extend Itself Even To Scientific Community? *Because though Hate Itself cannot not be Quantified ! Its certainly Quantifiable !*

As we shall see from the following scale I am proposing for these calculations :

K(The Ratio Of Hate)
=Number Of People Who Are Indoctrinated To Hate so they can kill /(Divided)By :
The Number Of People Who Are Killed By That Amount Of Hate :

$$K_P = \frac{G}{K.P.}$$

Where :

G is the genetic factor <u>Coagulated</u> after the period (P-In Years)?
And(K)Raised to the power of (P-Years)!

BOTTOM LINE.

"The above equation all that is telling us that the criminal genes are <u>Transmitted</u> According to the <u>Rapid</u> geometrical not arithmetical progression."

(See Page------Criminal Intent) ?

Exercise:

"(D X T = T X D) Is a <u>Mathematical Philosophical Field</u> where <u>Causality And Effects</u> Exchange places simultaneously ."

1- Can you <u>Superimpose</u> the above equation:

$$K = G$$
$$P \quad K.P.$$

On my master equation-5 of]: $D \times T = T \times D = Constant$.

2- Can you deduce the point of <u>Quantum Threshold</u> reached?

3- Why this <u>Quantum Threshold</u> Telling us that : Once the point is passed then the amount of hate though becomes <u>Invisible Its Irreversible</u> (Only for the <u>Protracting</u> Fascists i.e. the Colonialists)? ?

4- Is this <u>Irreversibility</u> Has anything to do with the <u>Irreversibility Of Time</u> i.e. <u>The Longer The Time (Protraction) The More Irreversible It Becomes</u>.

5- Does the resulting equations indicate <u>Geometrical</u> or <u>Arithmetical</u> Progression of the criminal genes ?

Also: (See Chapter Five-----The <u>BUVO</u> And The <u>EUVO</u>) ?

Matter Of Science.   Proving The Criminal Genes.

"There can be no JUNK DNA because nature create nothing in vaun").

"Although we do not know the <u>Absolute</u> numbers of criminal genes for each case ! We can by comparison calculate their <u>Relative Existence</u> ."

"The conscience of the colonialists may be dead but the <u>Subconscience Mind</u> never dies. Only to be stored genetically."   **The Red Monk.**

"Heaven and hell are here on earth not just in the life after: There is principle in the <u>Science Of Biology</u> that <u>Even</u> if an <u>Innocent</u> Child abuse their own body or get abused by others there will be <u>No Forgiveness</u> And that part of the body will eventually hit back in <u>Later Years</u> With related <u>Punishment</u> ! Similarly when a race like the English <u>Persist</u> (By Protractions) On committing atrocities against their fellow men (<u>Overtly Or Covertly</u>)Then these crimes will not just go away but are both <u>Stored And Spread</u> in what we call <u>Criminal Genes</u> only to come back haunting them later <u>Here On Earth Not In Life -After</u> !That is why you see inside the English speaking nations their ordinary daily lives are plagued with the miseries of crime and criminals !Stewing inside the <u>Hell</u> of en <u>mass Paranoia</u> Comes as standard in these crime infested nations. <u>Again You Do Not Need To Believe Anyone</u> :Just check it out for yourself : How Britain <u>By Far</u> the most <u>Destructive</u> and <u>Protracting</u> Compared to all other colonialists can now happily claim having <u>By Far</u> : the <u>Highest Rate Of Suicides</u> and the <u>Lowest Quality Of Life</u> ?"

Again You Do Not Need To Believe Anyone :Just Consider : Britain is the only country I know of where there is **Routine** Police presence outside every junior school in the morning and at closing times to prevent these kids from killing each other's As for the USA The frequent (**Shoot Out**)At their schools does not need my introduction .Crimes at such **Early Age** can only be attributed to the criminal genes going to action.__Let:__S=The Strength Of The Criminal Genes K=The Number Of Variations Of Crimes (In This Particular Example Its Five) G=Number Of Generations (Generation ≈ 50-70 Years).

Then We have : $S = K^G$

First Generation Criminal Genes.

Colonialism   Corruption.   Homosexuality Racism   Common Crimes

Second Generation Criminal Genes. Therefore in this example :
$S = 5^2$ For Second Generation. $S = 5^3$ For Third Generation.---And so forth.

PROOF:

Fortunately **ONE** of the proofs to the existence of **CRIMINAL GENES** comes from the science of **Psychoanalysis** which is **EASY TO UNDERSTAND:** Observe how the colonialist Europeans can not rest until they one way or another they **CRIMINALIZE** Those living among them who had came from their colonies?

REASON: As I kept saying at various spots of my books the human **SUBCONSCIOUS** never dead it sends all those crimes committed by the colonialists **GENETICALLY** ! Therefore **CRIMINALIZING** (WOGS)Is one way to **RELIEVE** this **Guilt Complex** (i.e. Justifying the crime they had already committed inside the colonies against these (WOGS) Who are after all nothing but (Shoplifters)! And more recently (Terrorists)Infact some western countries are known to finance train and send(Some Wogs)To commit acts of terrorismand the more atrocious these acts the better the coloniats feel!*(Search---European Colonialists In DAESH Robes)* ? The Subcoscious Mind .

"In sponsoring terrorism the dirty European Coloniasts are not only seeking material or military objective but also seeking urgent relive to their guilt complex by proving that (wogs)Are bad people deserving all that had happened to them."

Again to simplify with brevity this **Complex Subject:**
Everyone on this planet raised on the concept that killing or murder is

CRIME. Soldiers killing or harming wogs en masse under orders or under any make shift believe culture e.g. **PATRIOTISM** etc. Will do the evil deed without the approval of their subconscious mind .i.e. Priority is given to the Conscience mind that is why the other called SUB-conscience.But it does not mean the subconscious mind is gone away

The Price Of Genetic Mutations.

"Nature create nothing in vain including the misunderstood junk-DNA"

**In its (2024)Annual report the American Institute :
(None Profit Organization For Index of The Most Miserable Countries In the World).Its reports based on scientific studies !Interviewing REAL (Not Fake)People) Complex Statistical analysis of distress and clinically depressed !etc?etc??Uzbekistan (In Central Asia)Came out first as the most miserable nation on earth.But guess who came second ? ??(Great)Britain The number two most miserable country in the whole world c While we can understand why Uzbekistan miserable (Barren Terrain? Impoverished! etc!etc.)For Britain to end up second After looting half the world (Exporting Democracy) is mind blowing phenomenon!Not easily understood by most of us !As explained (Under the title Criminal Genes)in my books : Due to the protracting nature of British crimes against humanity these criminals GENES were given the time to mutate ; You see the conscience mind can turn blind eye to crimes committed in the name of Queen Country and state security but the Subconscious mind do not recognise these excuses .it record these crimes against humanity in what scientists used to think JUNK-DNA. Then only to be mutated producing the second most miserable country in the world.(Search-----The Proof) ?**(Search-----What Is Genetic Mutation) ?) ?

We already discussed under the titles of <u>Back Tracking</u> and (<u>Quantifiably</u>) That not everything can be quantified directly. In science often indirect methods are applied to measure the observable :Be it the <u>Invisible Atom</u> or the remote <u>Visible Stars</u> It's still <u>Exact Science</u>. We also quoted at various sections of this book that: By <u>Going As Far Back As Possible In History That Is As Far Back As Recorded Well Documented History Can Allow</u> : You will be surprised how accurate and well verified these collected numbers can be ? For example by comparing these <u>Real Figures</u> and <u>Concrete Numbers</u> it was discovered that the <u>Ratio</u> Of the <u>Total Number</u> of people killed or disabled by the British is at least <u>Twelve Times</u> the total number of people killed or maimed by the Germans. We also discovered to our <u>Ultimate Surprise</u> That this <u>Ratio</u> (Not The Absolute Numbers) Remained constant <u>Even</u> during <u>Peacetimes</u>!

A constant that is stubbornly refusing to budge ? !

Invariably pointing out to the solid existence of certain <u>Natural Forces</u> !

Insisting that be it <u>War Or Peace</u> Or any other conditions in between :

There are :

<u>Permanent Motivating Forces</u> Carried and driven by the most fundamentals inside our beings? i.e.

By biophysical-s such as <u>Criminal Genes</u>.

Which can only be permanently maintained by <u>Bio-Forces.</u>

Again by <u>Comparisons :</u>

We can obtain the <u>Exact Mathematical Ratio</u> Of these C.G. Though not their absolute numbers as follows:

Total number of people killed by the British at: *War times* ÷(Divided)By *Total* number of people killed by the Germans at *Wartimes*:=Equal

Total number of (Wogs) *Killed Or Traumatized* by the British at *Peace Times* (**See Pages---Silent Genocides**)? ÷ (*Divided*)By the *Total* number of (Auslanders)Harmed by the Germans at *Peace Times*:=(Equal)

The *Rate*(Per Capita)of common crimes committed inside countries of similar *Economic Systems* e.g. Britain or USA ÷(Divided)By the *Rate* (Per Capita)Of common crimes committed inside a nation of similar economic system e.g. Germany = (Equal):

The number of *Criminal Genes* Carried by the *Average* E.S.P. ÷(Divided)By the number of *Criminal Genes* carried by the *Average* German =8 to 12.6

Important Notes:

1- All Of These <u>Totals</u> Are Measured Over Equal Periods Of TimeSpans For <u>Each Case Indicated</u> Going As Far Back As <u>Concrete Indisputable</u> Records Of History Permits.

2- When we talk about the (<u>Rate</u>)Of common crimes we meanthe <u>Rate</u> of the total crimes divided by the population of the indicated nation !Or what is known as (Per Capita –Per Head)! Then we only begin to understand the horrors presented by these equations when we plug the <u>Concrete Numbers</u> !

For example inside the USA alone the American police shot dead three hundred sixty (360)<u>Unarmed</u> Black persons during the <u>first five months of (2015)Alone</u>! ? These are only crimes committed by the police themselves andwhen you add to it crimes committed by <u>None Police</u> the rate for the E.S.P will become unbelievably massive compared to those of <u>Modern Germany</u> Where the police had shot no one during those five months of-2015(There was only one single shooting incident in the <u>Past Fifty Years</u> !)As for the <u>Rate</u> of common crimes by <u>None Police</u> in Germany Its minimal.

Germany Was Chosen For Comparison:As Worst Case Scenario!

How The Criminal Gene Is Fed? And Why Its Growing??

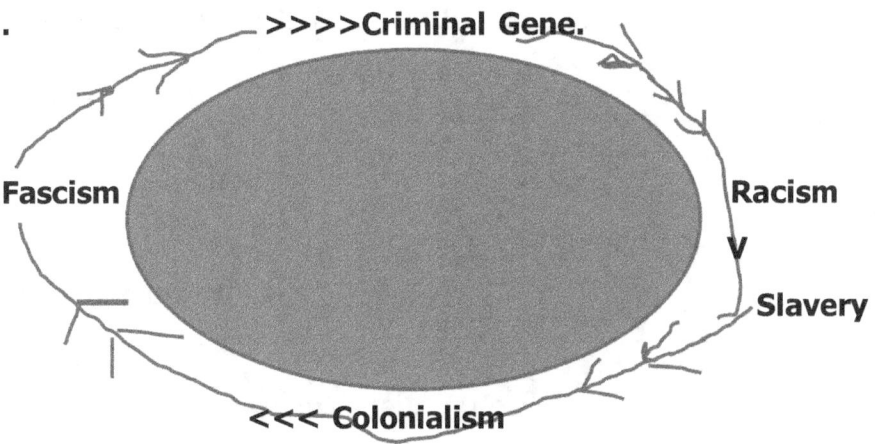

(See Page-----Matter Of Science) ?

(See Pages---The Criminal Genes Parts 1-72) ?

Relative Sizes Of The Blue Discs Represent The Relative Percentages Of Criminal Genes Present In Each Nation:

Criminal Genes

Fascism   **Racism (One Unit)**
(One Unit)

                              Chattel Slavery.

                              One Unit(The Minimum)

Colonialism(One Unit)

The British Case Average Criminal Genes (C.G) Fed By:

C.G = Racism + Chattel Slavery + Colonialism + Fascism = Four Units ➔ in real (3)Dimension world= 4^3 = 64%

The French Case Average Criminal Genes (C.G) Fed By

C.G = Racism + Colonialism + Fascism = Three Units ➔ in real (3)Dimension world of the DNA= 3^3 = 27%

The German Case Average Criminal Genes (C.G) Fed B

C.G = Racism + Fascism = Two Units ➔ in real (3)Dimension world of the DNA = 2^3 = 8%

The Compulsive Genes .

"The <u>Direct Proof</u> To the existence of these criminal genes is not by the horrendous facts and figures described in concrete terms by the above equations! Not <u>Even</u> in their Frequency <u>Or To That Matter Anything Else!</u> But found in their <u>Compulsive Nature</u>.
Their <u>Persistent</u> pattern surfacing with <u>Constant Rates</u> and <u>Inescapable Ratios</u> By these equations under :
<u>all conditions</u> and by <u>Any Argument Or Standard</u> .
<u>The Only Required Condition For These Rates And Ratios To Remain Persistently Constant Is To Be Among The E.S.P.Meaning It Can Only Be Genetic</u>. Period."

"What we need is to identify not only the <u>Frequency</u> of these crimes but also the <u>Compulsion</u> Behind all these <u>Unnecessary</u> Crimes with scientific methods which <u>Eliminate</u> Any other driving motives or forces from the argument except those of the criminal genes."

How it manifest itself right across the social board with <u>*Persisting Constant Ratio Under All Conditions*</u> *?In peace times and in war. In the conditions of common crimes committed by ordinary people who are not paid trained or meant to protect life and property ! As well as by those we call the police who are trained paid and meant to protect life and property !Or under any other possible conditions !{{During the first five months of (2015) Alone :*

The American police had shot dead three hundred and sixty (360)<u>*Unarmed*</u> *Black people (Official Figures)And the British police is not any different at all except these prefer the* <u>*Silence*</u> *of* <u>*Poisons ! Tasers!Or By Car Accidentsas As Well As Bullets!*</u> *Whichever* <u>*Untraceable*</u> **(See Page ---The Variety Principle)** *?Also* **(See Pages ---Silent Genocides Parts 1,2,Etc)** *?Their Victims include children or* <u>*Even*</u> *Elderly people over eightyyears old !(Official Statistics)* ! *Strikingly the (Rate) Of victims of* crimes committed by the British police themselves is **<u>Phenomenally</u>** of similar order of magnitude as that inside the USA in spite of the huge differences between their conditions ! ? ! }}

<u>What We Need Is To Identify The Compulsion Behind All These Unnecessary Crimes. By Scientific Methods That Eliminate Any Other Driving Motives Or Forces Except Those Of The Criminal Genes.To Achieve This</u>:

Let us award ourselves few moments of reflection on these figures by asking yourself :

Can I shoot a cat to death just because it was black and no one is looking (No Witnesses)?The answer is clearly (NO).That is :

<u>If You Happened To Be Not One Of Carrier Of Criminal Genes</u> . However carriers of the criminal genes inside Britain and USA Even when they are <u>**Trained Paid And Meant**</u> to protect life and property like their police :Find it <u>*Irresistible*</u> Not to commit crimes including the crime of murdering anyone they happened not to like

e.g. Shooting to death <u>Unarmed</u> Black people as long as they reckon no one is watching (No Witnesses)As it had been proven (360)Times in the first five months of (2015) ?

This is how <u>Compulsive</u> these criminal genes can be ?

And this is the <u>Ultimate Proof</u> To their existence ?

They are as <u>Compulsive</u> as the <u>Alcoholic Genes</u> Carried by <u>Most</u> Of the Irish.

Period.

"Just ask yourself : Are you able to shoot a cat to death just because it was black and <u>No One Is Looking</u> ? Clearly –No !But the carriers of criminal genes can and will ! Moreover they do not even know why??"

(See Page---Clay Pigeons)?

(See Page--Killer Race Parts 1-27) ?

BOTTOM LINE .

"Do not let Yourself ! Myself !Or Themselves !Be the judge . Let only these equations of Facts ! Figures !And History judge this race of criminals and faggots."

Whenever you come across a member of the E.S.P. Trying to defend or find excuses for their daily crimes against humanity :

Do not argue back !

<u>You Will Be Wasting Your Time Since They Are Driven Like I Said Purely By Genetics. Meaning They Not Even Aware Of It Or Do Not Wish To Be Aware</u> !!

Just <u>Add To The Equations</u> Stated above the daily number of people they <u>Kill? Maim? Or Traumatize</u> ???

And do not let <u>Yourself? Myself ? Or Themselves</u> Be the judge

Let only these equations of <u>Facts !Figures! And History</u> Judge this race of criminals and faggots.

4- The Red Monk Law .

" *Again* we apply here the scientific methods of backtracking."

"From the beginning mankind was an accidental birth in pain to live in pain! It's the duty of all politics and religions to act as doctors lessening the pain whenever and wherever they can but this does not mean abandoning the very set of rules that was evolved for this very purpose"

The Red Monk.

"When there are no values? The vacuum is filled by the culture of *Crime And Punishment*. Inside every corner of their life !(See Page----The Kojak Culture) ?

Defining The (U.V.M)Unit Value Of Morality.

Just as pound coin inside your pocket had travelled from one hand to another !
And from one generation to the next as _Monetary Unit_ !
And just as the pound grows in time (Yields) In to _Three Pounds_ (Due to inflation or otherwise e.g. investments) The units value of ethics can also synthesise (See Page---No Value No Synthesis)?
There are also _Units Of Morality_ That was coined (For _Example_) By the catholic church then passed from one age to the next until it re-emerged at the other side of the French revolution as the three universal values of : (Fraternity ,Liberty, Equality!).
Therefore a nation like Britain of zero values is considered to be _Morally Bankrupt_ A condition often precede _Material Bankruptcy._ Therefore we define the:

_U.V.M_ = Constant /Divided by units of _Immorality_ e.g. the concrete number of crimes per unit population. "

Since we cannot measure directly morality (M) But we certainly can measure immorality e.g. the number of people killed where;

M=1/(Divided By)the number of people killed.

Now keeping in mind that we are discussing here <u>Communal Morality</u> Of which <u>Individual Morality</u> is only one of its <u>Derivatives</u> And if we denote materialism by (m)Which is unlike morality its certainly quantifiable:

Then we have:

dm/dt = +ve.

m = for material.   t = for time.

dM/dt = -ve.

M = <u>Average</u> (Not Individual) Moral values.

We have already stated that we cannot measure morality (M) But we can measure immorality e.g. the number of overseas students destroyed by the Brits).

Then simply morality:

M=1/The Immorality.

HINTS.

What is $dM/dm =?$ Positive or Negative ??

What does the word <u>Negate</u> in the section above (w.r.t)

What the lag or phase angle "ɸ") means ? ?
Can you deduce [$e^{i\phi}$]   from the above ?
<u>*Plugging in some figures in to the equation :*</u>
Let Θ_i =Reaching the moon (Energy spent equivalent to that of one

atom bomb !Half mega ton of TNT.).
Θ_o = Killing one million Vietnamese,(just think of the <u>Inhumanity</u>?

How immoral And <u>Unnecessary</u> it was to deploy <u>Agent O.R.A.N.G.E</u>

Then: $\Phi = \Theta_o - \Theta_i$ = Theta Output –Theta Input .

It's essential for our definition to note the condition of :
<u>(Zero Reasons)</u> Or <u>(Not Necessary</u>)It was to apply the <u>Defoliage</u>

<u>*Agent Orange*</u>*. A defoliage not only still killing God's greens but also human beings!* **(See Pages ------------------- Killer Race) ?Also :**

(See Page---American Sniper) ?

The Mathematics Of Hypocrisy.

"This e following are just st basic foundation for further calculations based on set thorny "

"Here the ACCUMI"LATION of Dislocation of MICRO judgments offer goodillustration of how the material (e.g. Quality of Life)And the Non Material(e,g, Hypocrisy) Are Organically linked ."

"Rule of thumb: The less there are Misjudgments in any sphere of life by the state or individual the better will be the quality of life and vice versa)."

Let A , B, and C Forming REAL SET of (S)Representing thre Real persons inside any community where IPurpose built and made to measure ies lying and dirty constructions cpmes as standard considered to be clever by their

(Commoners)And Acts of clever patriotism by their Medeival Ruling Circles considered to be clever by their (Commoners)And clever acts of patriotism by their medieval ruling circles

When A pretend he is A* Then B & C Subconsciencly reciprocate by pretending with the the minimum (Minimum means high ist probability for such reactions by C & D.

Hence we have the image or pretended st set (S)* :

Now let the judgment based on the realiy by (s)= J(S) and that :

based on the pretention of (S)*= (JS)*

$$J(s) -(JS)* = -(MINUS)\ \delta j$$

Therefore the difference between the two judgments =The SMALL Misjudgment (δJ)

As δj →d J.

$$F(q) = 1/\sqrt{2}\ \Pi \qquad R$$

Here we introduced another function (q)For quality of life:Clearly no misjudgement can ever lead to improve the quality of life because simply it will not be called misjudgemebts.

The limits of integration are between the (*N)For imaginary or pretended state while thr R(Foe reality)

Clearly the deuteriation of quality of life due to hypocracy is evident from the above mathematical expression

This result is fully backed by <u>Practica</u>l observation of societies (UNDER COMPARABLE CONDITIONS)the less hypocritical the better quality of life

For qualitative definition of (q)Can be found in any scientific Socio-Economic text book:

(Search My E BOOKS ------The limits of hypocrisy Parts 1-5?
Foe how so many Europeans Visting Britain that i met expressing their shock at the rate of DETERIATION of QUALITY Of LIFE And latest United Nation ALARMING Report on POVERTY inside BRITAIN? It's the PRICE of British Institutionalized Hypocrisy.

BOTTOM LINE.

1-In this Frequmechanics -Three we established

Very important equation that can identify Digitally every object in the universe MICROSCOPIC OR MACROSCOPIC Relative to the Cosmic Flux. As points in the cosmic fabric of space.
Based on the Phase or Lag angle between two frequencies cutting and rejoining the fabrics of the cosmic flux.
For this we derived the following equation.
although (Alpha ($\overset{7}{\alpha}$))= $10^{-(Minus\ Betta)}$
The range of $\beta = 0 - \infty$ Thus the region will be a MAP from ($œ:\beta$)
$œ = (1-\ 0)$ To Betta $(0-\infty)$
Such that: $(œ)\mu\ = (H)\mu, V\ \beta v$
Where (H) Is some arbitrary regional Constant.

2-The Limiting Limitations Of Nature:

"Nature never reveals its own secrets fully."

a-At the beginning of this book we asked if Einstein was aware that NATURE had IMPOSED other limits on our knowledge apart from the limits of speed of light.

b-And it was established in this book how the phase angle too is LIMITED i.e although Φ T h e P h a s e a n g l e i s :
(0-1) And its RANGE (0-∞) it not allowed to reach ONE regardless of how TECHNICALLY precise its FEEDBACK;
If it does then the two frequencies are exactly equal it mean there is only one frequency (Ω)Thus there is NO COSMIC FLUX to cut and to Re- join,

c-
Another limits IMPOSED by nature is the CONSTAN OF DESTRUCTIVY which we calculated by VALID COMPARISON of two European THUGS (British Colonialists and German Fascists) By calculating the DISTRUCTIBILITY of each discovering two exactly

equal areas confirming there is a constant of destructivity PER ISSUE PER Two COMPARABLE Entities. As shown by the Equations and the Diagrams above.

d-Another limits imposed by NATURE is the very Principle that Quantum Mechanic is built upon which is (We cannot see or measure two events Simultaneously! God does not play dice (Einstein)But HE certainly limited our dice in to SIX FACES?

This is true of the microscopic world as well as the macroscopic where many of the lights we observe in the sky belong to era of billions years age i.e. LAG Angle of Billion years.

In conclusion its reasonable to assume there is ORGANIC as well as INTRINSIC LINK between these FOUR limitations imposed by nature on the speed of light !the phase Angle! And the self-destructibility of Man .And I go further by claiming that these four limitations are

PHYSICALLY RELATED CONNECTED and Reciprocated.

(Search My Other Books For Neuther Current-Physics)?

In Search For My Universal Constant (H^n) Part Seven.

"Its not easy to Quantize the macroscopic world ! Will the answer be inside Multiplets of Planck Constant (n.h)? If its valid for the photon ? Why not for othe particles??"

My equation in its basic form:

$$(\Omega) = (\omega\text{cutting}) \pm (J\omega\text{Gluing})$$

$$(\Omega)^2 = [(\omega\text{cutting}) + (J\omega\text{Gluing})]^2 \rightarrow (\Omega)^2 = [(\omega c)^2 \; (\omega_g)^2 \pm 2J \; \omega c \cdot (\omega_g)]$$

Under the title of introducing the cosmic flux (Δ) Established:

$$\Omega^n = (\omega c)^n - (\omega G)^n + J.\Sigma \, (n!).(\Phi)r /(n-r)!r$$

$$(H.\Omega)^n = (h.\omega c)^n - (h.\omega g)^n + J.\Sigma \, (n!).(\Phi)r/(n-r)!$$

The Main Leading Real Terms are :

$$e^{(H.\Omega)n} \approx e^{(h\,\omega c)n} \cdot e^{(h\cdot\omega g)n}$$

Differentiating w.r.t (ϖ)(for definition of (ϖ) See The Isamic Wave ?)Due to computer restrictions beware of displaced positions of powers e.g(Minus n)!

For now, conveniently **Exchange** the powers (n) By another:

n=Number of repeated Diff operations?

$$d^n/d\varpi \; e^{(H.\Omega)n} \approx d^n/d\varpi \; \{e^{(h.\omega c)n} \cdot e^{(h\,\omega g)}\}$$

$$d^n/d\varpi \; e^{(H.\Omega)n} \approx e^{\underline{(h\,\omega g)n}} d^n/d\varpi \; \{e^{(h.\omega c)n} + e^{(h.\omega c)n} \; d^n/d\varpi \; e^{(h\,\omega g)n}$$

$$\approx h^n . e^{(h.\omega c)n} \cdot e^{(h\,\omega g)n} + (h)^n \cdot e^{(\omega g)n.h} \; e^{(h.\omega c)n}$$

$$H^n . e^{(H.\Omega)n} = e^{(\omega c +\omega g)n.h} \; \{h^n + \delta n \; (h)^s\} \qquad \text{Then at } \Omega = 0$$

$$H^n = (\tfrac{1}{2})[h^n + (h)^n] . e^{\delta n (h.\omega c + h\,\omega g)n}$$

$$H^n = (\tfrac{1}{2})[h^n + \delta n.(h)^{\delta m}] . e^{(\omega c + \omega g).S}$$

Introducing My Fifth and Sixth Isamic Deltas *(δn & δm)*

Now for ($\approx$) → (=)

Let: $S = (n.h)^{\delta m}$

At $n=0$? $\delta n = 1$ and

at $n=1$? $\delta n = 0$

At $n > 1$ $\delta n = +1$. AND:

$\delta m = -n$ $S \rightarrow (n.h)^{-n}$

Giving: $H^n = (½)[h^n + \delta n.(h)^{\delta m}].e^{(\omega c + \omega g).S}$

$H^n = (½)[h^n + \delta n.(n.h)^{-n}].e^{(\omega c + \omega g).S}$

h = Planck Constant (erg.Sec) !

H = My Proposed Universal Constant (erg.Sec).

Ω = Hybrid Frequency made of two frequencies ωc & ωg!

n = 1, 2, 3,..etc

ϕ = <u>Relativistically</u> calculated Phase angle By:

-Any two observers at two different locations inside same region of (Roberson-Walker)universe <u>observing</u> Ω By measuring <u>Relativistically</u> angular velocity Omega telda (Telda (Ω_R)).

2-

By one single observer of Both frequencies ωc and ωg by measuring <u>Relativistically</u> the two angular velocities Omega Telda ($\tilde{\omega}_{Cutting.}$) And ($\tilde{\omega}_{Gluing.}$).

<u>BOTTOM LINE.</u>

"Any Absolute platform of assumptions automatically yield. absolute mathematical valueumbers).

Land Of Plenty! Extracts From My Principle Of Prolificsy:

The Philosophy:

"The ultimate destiny of any increasing QUANTITY will be QUALITY (Turning Discrete);" **The Red Monk.**

The Science:

"My principle states that: If there are plenty enough numbers of Exact Copies of any Macroscopic objects then it should display and obey quantum effects and rules."

" Observe the pattern of ripples of yhe high seas and the wave mechanics offered by Sea Waves !Observe the strange topology of the sand waves in (AL-AHQAF) Also known as the (Empty Quarter) A region the size of big country South east of the Arabian Peninsula Covered entirely by pure fine Identical Particles of sands !No one has ever dared entering and had come out alive !Its Unexplained Patterns of behaviour mentioned several times even in the Holy Quraan."

$$N/Z_{(A)} = K. = 1/500 \times 10^{+27}$$

Field Of View :

"The reason why we cannot observe quantum effects in macroscopic objects :Because according to my Prolificacy Principle there are just not enough Exact copies falling inside our field of view."

Let $r(A)$ = The radius of the smallest atom that display quantum effects singularly.

$R(A)$ = The radius of the macroscopic objects .

n = The number (Whole Digits) of copies of the Macroscopic in any ensemble required in our field of view before we can observe any quantum effects then

$[R/r]^n$ = the minimum number of copies necessary for any observable ensemble displaying quantum effects.

Clearly this shows how the size of the *Macroscopic* object multiplies shooting up? While the *Microscopic* size of the standard atom is miniaturized by te power $(1/n)$

!! (See Specific Infinities)?

Let observer (O1) At distance (S1) From (n1) Number of observables ! And observer (O2) At distance (S2) From (n2) Minimum Number of observables then As first approximation n1>n2 And S1 > S2 Let S1-S2 =ds And n1-n2=N

n1-n2 =K.(e^{S1} -e^{S2})=N Multiplying through by (e^{-s2})

N . (e^{-s2})=K(e^{ds} -1) Clearly at ds=0

e^{-s2} =(K/N)(e^{ds} -1)=0 therfore S2 → ∞

Translated (The closer the distace between the two observers become the more infinite their field of view) Now Dividing :

Let R= Ratio of n1/n2 R=K.(e^{S1} /e^{S2})= K.(e^{S1-S2})= K.e^{ds}

$$dS = \log R/K = C.dt \text{--------------------} EQN\text{-}M$$

Let (n1-n2)=K.(e^{S1} -e^{S2}) , n1/n2= K.e^{ds} Subsituting for n1 =n2. K.e^{ds} n1-n2 = n2. K.e^{ds} - n2

Dn=n2 (ke^{ds} -1) Giving :Dn/n2 =(ke^{ds} -1) !

Again implying that as the observers get closer and ds→0 the ratio :(n1-n2)/n2 = (R-1)→(K-1) i.e. R=Constant!

Just another confirmation that speed of light is Constant. C=Speed of light A.Z =taken to be Constant of the locality. We Have Now Extablished Tool For Measuring The Distance (ds) By The Difference IFrequencies ($\delta^c{}_g$)

By EQN-M: $ds/dt = C$. Keeping in mind these are only the <u>Extracts</u> from later main section ?(Prolifcasy Principle)?and that : n=Number of microscopic inside the macroscopic but:

n1 And n2 Are the numbers of observables! Therefore

n1 −n2 = The difference in the numbers of observables by O1 and O2 at distances of S1 and S2 ! **(See Specific Infinities)?**

IF: $(Z/A)_1 = (Z/A)_2 = n$ then from my <u>Master Equation</u> in tensorial form: $\Omega_R{}^{\mu\cdot\nu} = (\tilde{\omega}_c)^\mu \pm j \cdot (\tilde{\omega}_g)^\nu$ V space → V* space

$n \cdot \Omega_R{}^{\mu\cdot\nu} = n_1(\tilde{\omega}_c)^\mu + j \cdot n_2 (\tilde{\omega}_g)^\nu$ IF:

$(Z/A)_1 \neq (Z/A)_2$ *Then:*

$<_{(Z/A)_1} \big| \Omega_R{}^{\mu\cdot\nu} \big|_{(Z/A)_2}> \; = n_1(\tilde{\omega}_c)^\mu + j \cdot n_2 (\tilde{\omega}_g)^\nu$

$(\infty)_{n_1} - (\infty)_{n_2} = \delta_\infty$

$\delta_\infty = f\,[1/(n_1) - (1/n_2)\,]$ At $n_1 = n_2 = 0$ $\delta_\infty = \infty$

And At : $n_1 = n_2 \neq 0$ $\delta_\infty = 0$ Since at $\omega_c = \omega_g \delta^c{}_g = 0$

Put : $n_1 = Z_{(C)} \cdot \omega_c$

$n_2 = Z_{(g)} \cdot \omega_g$ hence:

$R = \phi$ At $Z_{(C)}$ $= Z_{(g)}$. And proceed accordingly?

(See Specific Infinities)?

www.ingramcontent.com/pod-product-compliance
Lightning Source LLC
Chambersburg PA
CBHW071350210526
45465CB00001B/45